The Face of Florida

Third Edition

Clarence M. Head
University of Central Florida
Orlando, Florida
Robert B. Marcus

KENDALL/HUNT PUBLISHING COMPANY
4050 Westmark Drive Dubuque, Iowa 52002

ISBN 13: 978-0-7872-4849-9

Printed in the United States of America

10 9 8 7

Contents

Preface

This book has evolved over a period of years through the teaching and research of the two authors. At the University of Central Florida Professor C. M. Head uses it as part of a larger course in Physical Geography, while at the University of Florida Professor R. B. Marcus teaches it as *The Face of Florida.* It is also offered as a correspondence course under the same title.

The material covered in the book is intended to provide the student with a knowledge of the special aspects and features of Florida's physical geography. Special attention is directed to Florida's paleogeography, as well as Florida's present landscape, climate, soils, and vegetation.

The book can be used in Physical Geography, Geology, Earth Sciences, and The Geography of Florida. More detail is given in several chapters than might be deemed necessary, but the detail is included to give a background to those students who have little acquaintance with the Geography of Geomorphology .

The book is intended for community college students as well as students in four-year colleges and universities. It is also appropriate for advanced high school students at the senior level.

In format the book is divided into eight chapters. Chapter 1 discusses the ways of locating a point or an area on the earth's surface. It also discusses how the federal land that comprised the territory of Florida was transferred from federal to state ownership. The chapter concludes with a description of the size and shape of the present-day State of Florida. In Chapter 2, Florida's features in the geologic past are traced to show the student the changes the area has undergone over long periods of time. Chapter 3 is devoted to the climate of the state with special emphasis on the effects of location upon climate. This chapter sets the stage for the next four chapters that are devoted to the physical geography of present-day Florida.

Chapter 4 depicts Florida as it is today while in Chapter 5 the processes beneath the surface of the state are discussed. In Chapter 6 the most rapidly changing feature of Florida, its coastlines, are examined both in the geomorphology context as well as the human context.

Chapters 7 and 8 are devoted to the soils and vegetation and include some of the problems and changes that have evolved as the vegetation is cleared, the soils disturbed, and virgin areas covered with asphalt and converted into urban communities to accommodate the rapidly increasing population.

It is expected that students and faculty will have valuable suggestions for further revisions. Any corrections and suggestions for revision are welcomed by the authors.

C.M. Head
R. B. Marcus

Aknowledgments

The authors wish to acknowledge several agencies and individuals who have given valuable assistance in the preparation of this book. The groups include the staff in the Library and Related Services at the University of Central Florida, the Florida Geologic Survey, the Florida Game and Fresh Water Commission, and Jammal and Associates, Inc.

Individually, James A. Henry and David J. Bynum were always willing to provide information we requested on climate and vegetation, respectively, sometimes devoting considerable time researching the information. E. C. Pirkle provided both information and suggestions on many occasions. Most of the maps and diagrams were drawn by Paul A. Stayert and Stephen H. Rogers of the Department of Geography, University of Florida and Michael Hanson at University of Central Florida.

Special thanks go to our wives, Jacqueline Head and Catherine Marcus, for their aid with the proof and for their many helpful suggestions. Thanks is also due to Mary Robinson who typed several chapters of the manuscript and who worked at night and on week-ends to complete the tasks.

We also wish to thank all of those unnamed students at both universities who indirectly have contributed, through dialogue and inspiration, to the completion of this book.

CHAPTER 1
Florida's Location, Shape and Evolution

INTRODUCTION

The study of Geography, whether physical, cultural or economic emphasizes spatial distribution. In this text the main emphasis is on physical so it is important to understand how features are distributed over an area. It is also important to understand the basic systems of superimposing a grid system over the earth. This makes it possible to locate exactly, by coordinates, a point or an area relative to another point or area on the surface of the earth. We call the former method of location **mathematical location** and the latter method **relative location**.

MATHEMATICAL LOCATION

Mathematical location is an exact method of determining the position of a point on a grid system superimposed upon the earth's surface. A particular point may be located by means of **latitude** and **longitude, range** and **township,** or some other set of coordinates.

Latitude is measured in a north-south direction from the **equator** (0°) along a meridian or a great circle. The north and south poles are 90° north and south of the equator respectively. The equator, which is midway between the two poles, can be located by observing the position of the sun on March 21 and September 21. On the two days the sun is directly overhead at the equator at noon and the points are referred to as the **equinoxes**.

Longitude is measured in an east-west direction along a parallel of latitude from the **Prime Meridian** (0°), an arbitrarily accepted imaginary line which extends from the north pole to the south pole and originally was placed at the intersection of the crosshairs on the Royal Observatory telescope at Greenwich, England. It is frequently referred to as the **Greenwich Meridian** or the Prime Meridian.

Points on the surface of the earth are measured north and south from the equator and east and west from the Prime Meridian in degrees, minutes, and seconds. By using such figures each point on the earth's surface has its own distinguishing value and can be easily located by knowing its particular latitude and longitude.

Another commonly used mathematical grid system is the range and township system. On May 20, 1785, the Continental Congress passed "An ordinance for ascertaining the mode of locating and disposing of lands in the Western Territory, and for other purposes therein mentioned."

The use of **range** and **township** is known as the rectangular system of surveying and marked an important transition from the surveying practice that characterized most of the area included in the original 13 colonies. Under that surveying practice land grants were defined by irregular metes and bounds, each depending upon a description of the adjoining tract, which was known by a name or survey number. These tracts usually lacked a common geographical location except for a reference to some well-known natural object.

The ordinance for rectangular survey passed by the Continental Congress states:

> The public lands shall be divided by north and south lines running according to the true meridians and by others crossing them at right angles to form townships of six square miles, unless where the line of an Indian Reservation or of lands heretofore surveyed or patented, or the course of navigable rivers, may render this impracticable; and in that case this rule must be departed from no further that such parcircumstances required.

The meridians, or "range lines," are cardinal lines (true meridians; great circles on the earth) running north and south. They form the east and west boundaries of a township. The principal meridian for the State of Florida (also for part of the State of Alabama) is located along a true meridian in Tallahassee at 84° 16′ 38″ West Longitude. Since "range lines" are not parallel and converge at the poles, each section does not contain 640 acres. The greatest variations are found in the sections along the north and west boundaries.

"Township lines" are parallels of latitude and mark the northern and southern boundaries of the townships. Since the equator is the only parallel of latitude that is a great circle, all other parallels, including those designated as "township lines" are small circles. The base line for the State of Florida is located in Tallahassee at 30° 26′ 03″ North Latitude (Figure 1-2).

The term township carries two connotations. The first refers to the "township lines" which form the northern and southern boundaries of the areas of land six by six miles. The second connotation applies to the area of land six miles by six miles or 36 square miles (Figure 1-1).

Each township is subdivided into 36 sections. A section is an area one mile by one mile and contains as nearly as possible 640 acres. Half sections contain approximately 320 acres and quarter sections approximately 160 acres respectively. The original land grants of public lands to Homesteaders was 160 acres. Sections are numbered consecutively from 1 to 36 beginning in the northeast corner of a township and continuing westward for six miles, then south one section and eastward for six sections back and forth until section 36 in the southeast corner of the township is reached (Figures 1-3 and 1-4).

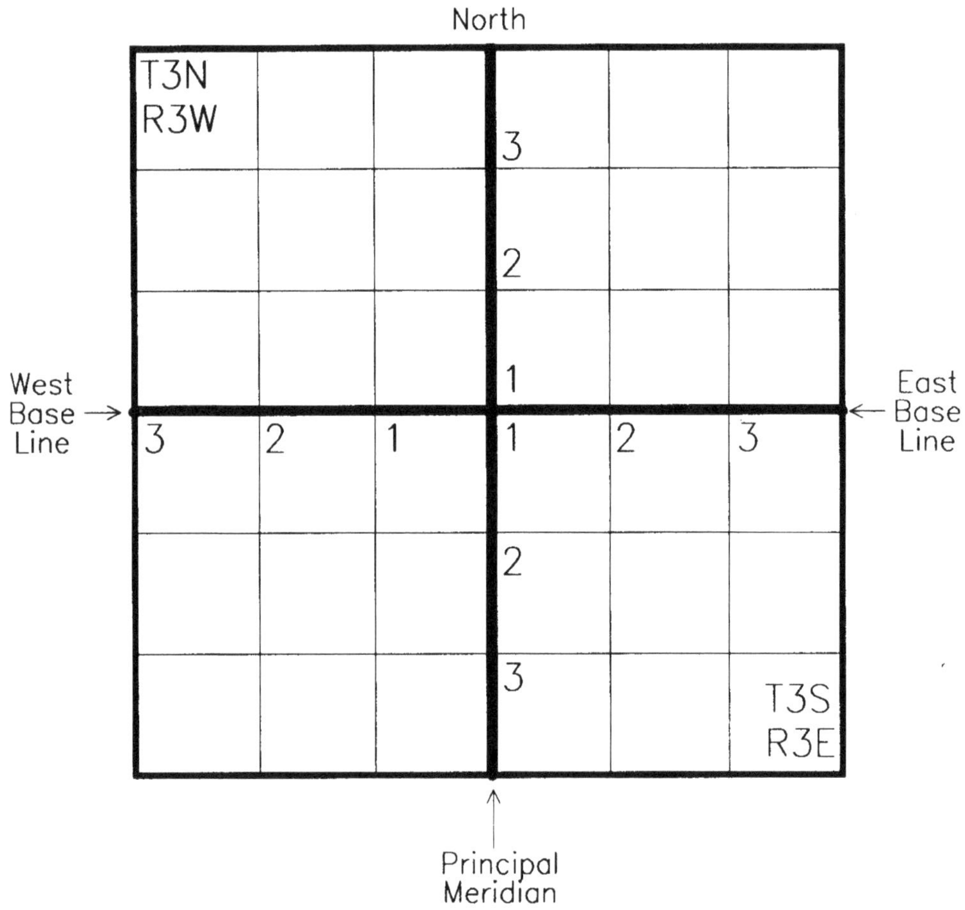

Figure 1-1. Range lines run parallel to the principal meridian. Township lines run parallel to the base line. Range lines and township lines delineate a township, six by six miles.

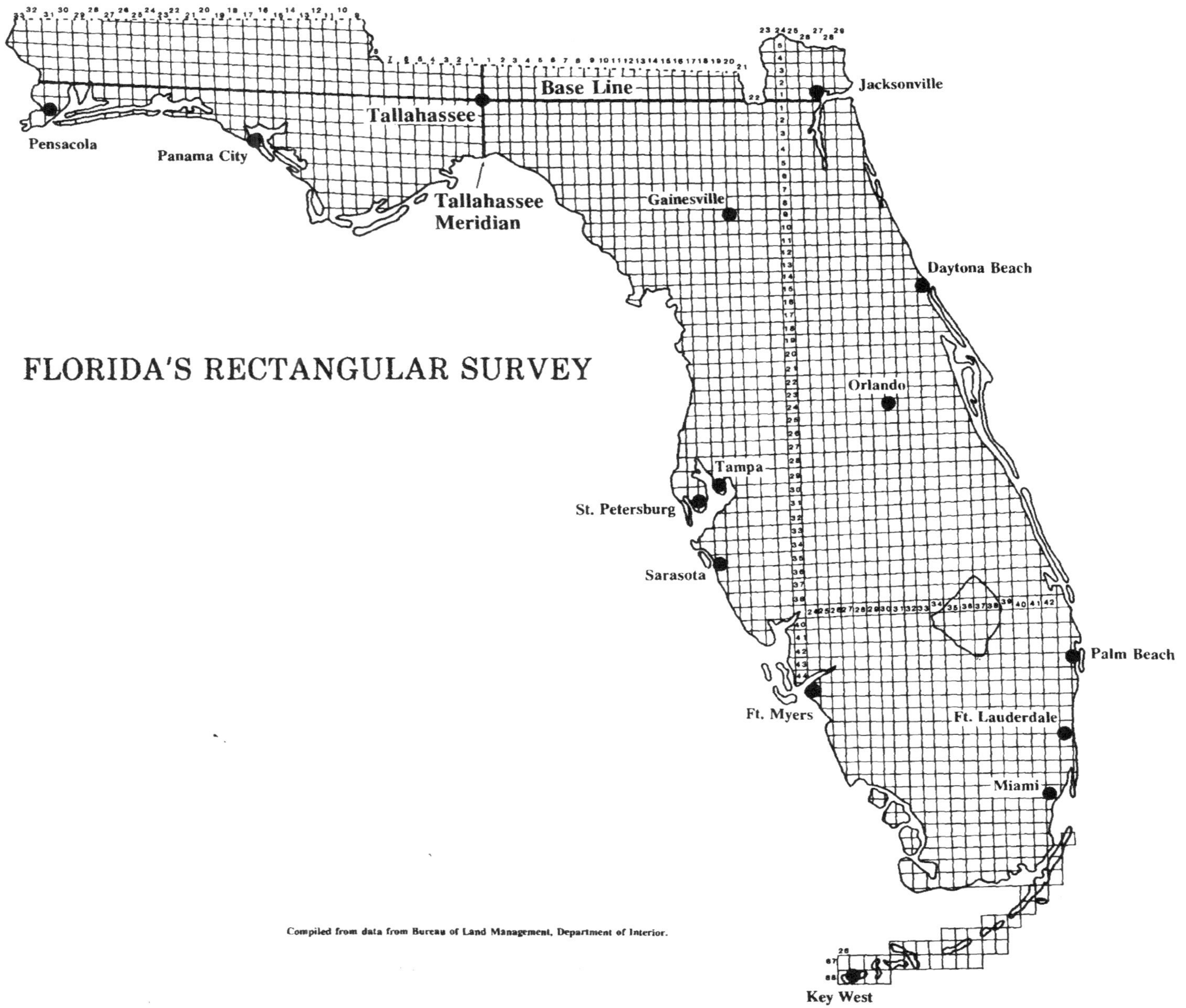

Figure 1.2. Florida as surveyed by rectangular survey. The original 13 colonies plus several other states were surveyed by metes and bounds. All townships are surveyed from the Tallahassee meridian and the base lines which intersect in the city of Tallahassee.

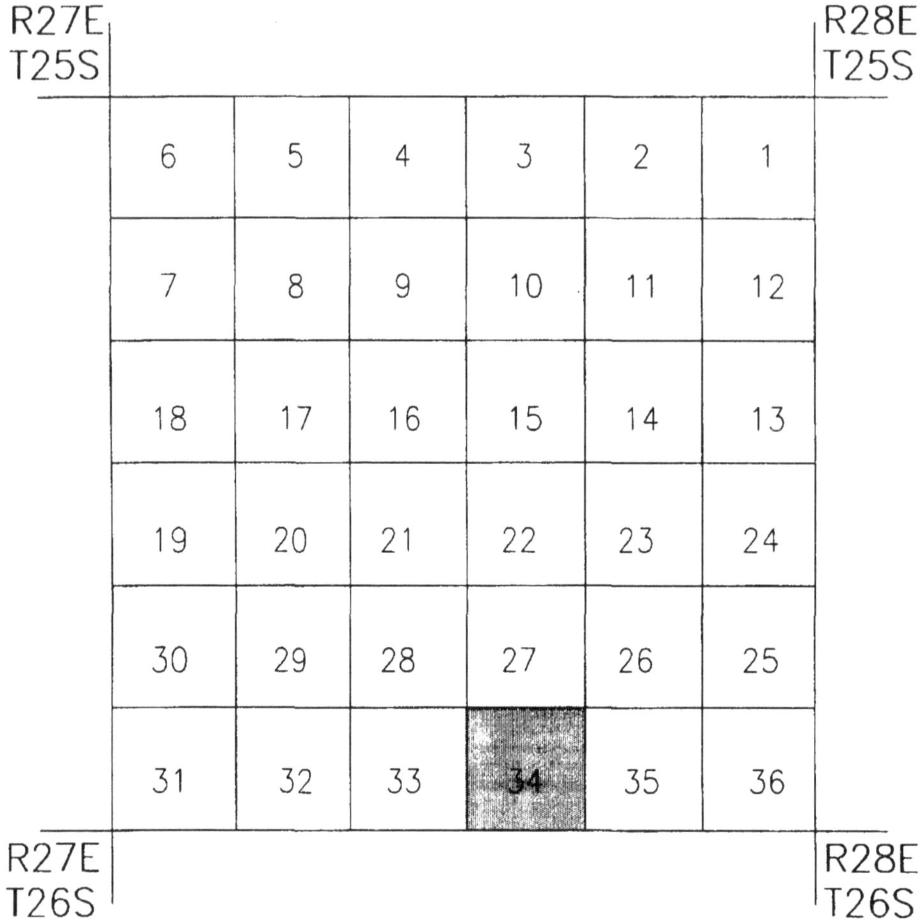

Figure 1-3. Each township is divided into 36 sections, each one mile by one mile. They are numbered beginning with NE section and ending with SE section.

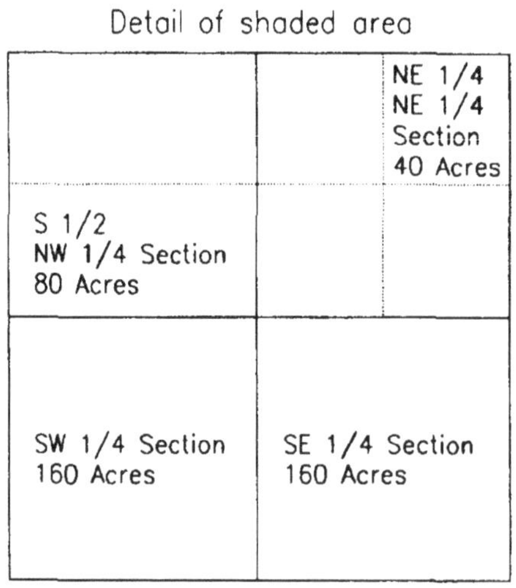

Figure 1-4. Each section may be divided into one-quarter sections, one-eighth sections and even into one-sixteenth sections.

RELATIVE LOCATION

The second method of locating a point or an area is relative. Relative location implies a location in generalized terms in relation to other points or areas on the earth. The comparison may be in reference to the cardinal directions of the compass; namely north, east, south, or west or in terms of some other generalized location, such as left or right, or above or below. Florida's latitudinal location, between approximately 24° 31′ North Latitude and 31° 00′ North Latitude, is important because this places the state within the latitudinal belt of the great deserts of the world. However, the great deserts located within this latitudinal belt are located on the west side of continents. Florida, despite its location within this latitudinal belt, is located on the east side of the North American continent and is, therefore, one of the wettest areas in the United States.

Another way to visualize relative location is latitudinally. Consider an automobile trip from Key West to a northern city, such as Cincinnati, Ohio or Washington, D. C. Almost half of the trip will be in Florida. Figure 1-5 shows another way to compare the size of Florida latitudinally with other areas of the United States. This map shows the State of Florida placed upon an outlined map of the eastern United States with the northwestern corner at Chicago, the northeastern border at Pittsburgh, and the southern tip of the peninsula extending to about Charlotte, North Carolina and the Keys reaching into South Carolina.

Florida's latitudinal location in relation to the United States extends to a favorable location in relation to Latin America. It is the closest state to the median points of both population and land masses of the Americas. This means that it is near the middle of all of the landmasses and the population center of the western hemisphere.

A study of longitudinal position shows that the easternmost point of the state, Palm Beach, is almost directly south of Pittsburgh, Pennsylvania while Pensacola located at the western end of the state is almost directly south of Chicago, Illinois. This relative location in the southeastern section of the United Section places the state in a favorable location with respect to about 60 percent of the nation's population. Many people who live in the colder regions of the United States, east of the Mississippi River, come to Florida for their winter vacations. They come because Florida is much closer to their homes than Arizona or California.

When NASA searched for a location to launch its rockets and space vehicles, Florida was selected because of its location on the eastern margin of the North American continent and its peninsular shape. Florida's longitudinal location provided two assets for launching space vehicles: (1) firing toward the east, the direction in which the earth turns, adds an additional 1,000 miles per hour (2,600 kilometers per hour) to the launch velocity and (2) the South Atlantic Flight Range extending eastward from Cape Canaveral provides a large unpopulated, easily controlled area for the recovery of or the destruction of malfunctioning rockets.

Today Florida's relative geographical location is an important asset to the state. However, this is a modern-day phenomena, that has largely developed since the turn of the last century. For almost 300 years Florida's location was important to Spain only because it lay along the sea routes of the Spanish galleons as they sailed from Mexico to Spain. It offered a haven for Spanish sailors who were shipwrecked along the Florida coasts. The attractive features, such as the mild winters and beautiful beaches, that attract so many today, were unimportant to the Spanish of the 16th, 17th, and 18th centuries.

PUBLIC DOMAIN

The 58,560 square miles, or approximately 35 million acres, of land and water that comprise the State of Florida was originally public domain. The term public domain applies to land whose title is subject to the administration, survey and transfer of title under the public laws of the United States and at present comes under the jurisdiction of the Director of the Bureau of Land Management, a subdivision of the Department of the Interior. Twenty-nine of the 48 contiguous mainland states are classified under this category. All of the 29 states were surveyed under the rectangular system of survey. The original 13 states plus the states of Maine, Vermont, West Virginia, Tennessee, Kentucky, and Texas retained title to all of the vacant land in the respective states and the United States public land laws are not applicable (Figure 1-6).

Four years prior to statehood, the Federal Government through a Federal Act of 1841 gave the Territorial Government of Florida 500,000 acres of swamp land to be used for internal improvements. Such improvements included roads, railroads, bridges, canals, and improvement of water courses. Money could be raised by the Territorial Government from the sale of the land, however, the minimum price was set at $1.25 per acre. This was the beginning of the change of title of public domain to the trustees acting for the State of Florida. The Federal Government could also sell land directly to individuals or groups through Federal Land Offices. During the Territorial Period the Federal Government sold 796,891 acres through the Tallahassee Land Office, 70,155 acres through the St. Augustine Land Office and 55,459 acres through the office in Newnansville (near present-day Alachua). (Dovell, 1952, Tebeau, 1971).

After Florida was admitted to the Union in 1845, the Federal Government continued to hold title to unappropriated lands and to administer the public-land laws. Congress alone has the power, derived from Article IV, Section 3 of the Constitution, to dispose of public domain and to make all rules and regulations that apply to the land.

In 17 of the 29 mainland states originally classified as public domain, the swamp and overflow lands, through public domain, became the property of the states when the lands were identified by public land survey. The title of the land then became subject to disposal by the states.

The Swamp Lands Act of 1850 gave the State of Florida title to 20.4 million acres of land. With the additional land acquired in 1841, Florida held title to 21 million of the 35 million acres lying within the state boundaries. The Swamp Lands Act of 1850 specifically required that the proceeds from the sale or direct appropriation of the land be applied exclusively for the reclamation of the land by building levees and draining the wetlands. The phobia for draining wetlands that has gripped many Florida developers over the past century had its beginning with this law. The last large block of public domain to be transferred to the state was included in Patent Number 137, issued April 24, 1903 and encompassed some 2.8 million acres of swamp and overflow land. The reason for the delay until 1903 resulted from the failure to have the land lying between Lake Okeechobee and the southwest corner of the state described by metes and bounds, the simplest way to fulfill the requirement that an area be surveyed before transfer of title could be accomplished.

At the beginning of the 1850's, Florida was a poor state. One of the few assets the state had was land. However, to attract settlers and dispose of land, transportation was sorely needed. Guidelines established by the Federal Government for the disposal of public domain included clauses permitting the state to either sell the land or to give it to railroad companies as an incentive to construct railroads.

An Act to Assist Railroad Construction in Florida was passed by the United States Congress in May 1856. The Act permitted the state to give alternate sections of land designated by odd section numbers within six miles of a proposed railroad on either side of the railroad. If any of the designated land had previously been sold by the Federal Government, equal acreage within 15 miles of the proposed railroad would be granted. The land was to be disposed of by the State of Florida in 20-mile segments and the title to the land would be transferred to the railroad company upon certification by the Governor to the Secretary of the Interior that the 20-mile section of railroad had been completed. For each 20-mile segment of railroad completed, it was possible for a railroad company to obtain title to 76,800 acres of land. At one period between 1870 and 1900, the railroads in Florida had

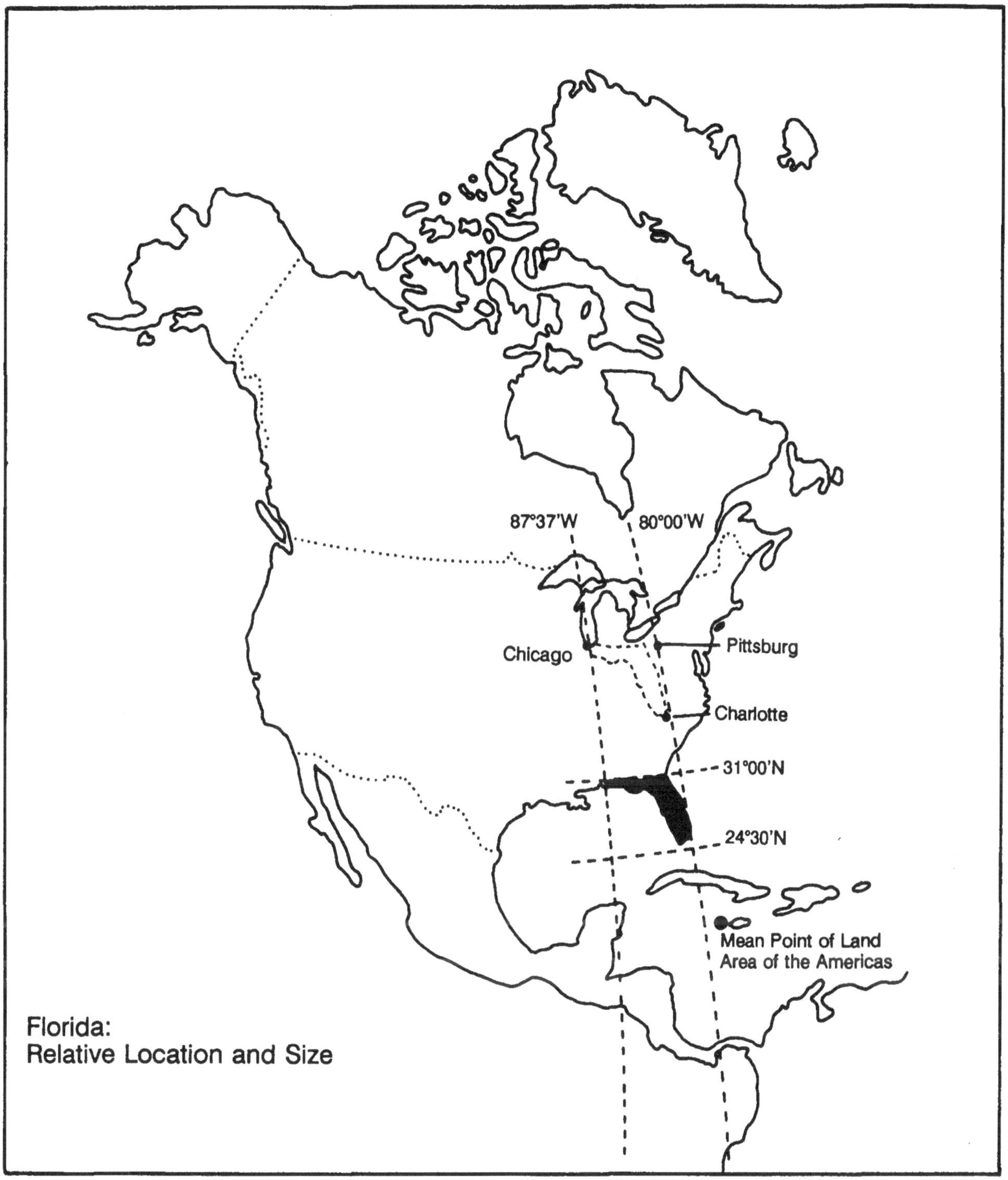

Figure 1-5. The relative location of Florida shows that longitudinally it extends from Pittsburgh to Chicago and latitudinally from Pittsburgh to Charlotte, North Carolina.

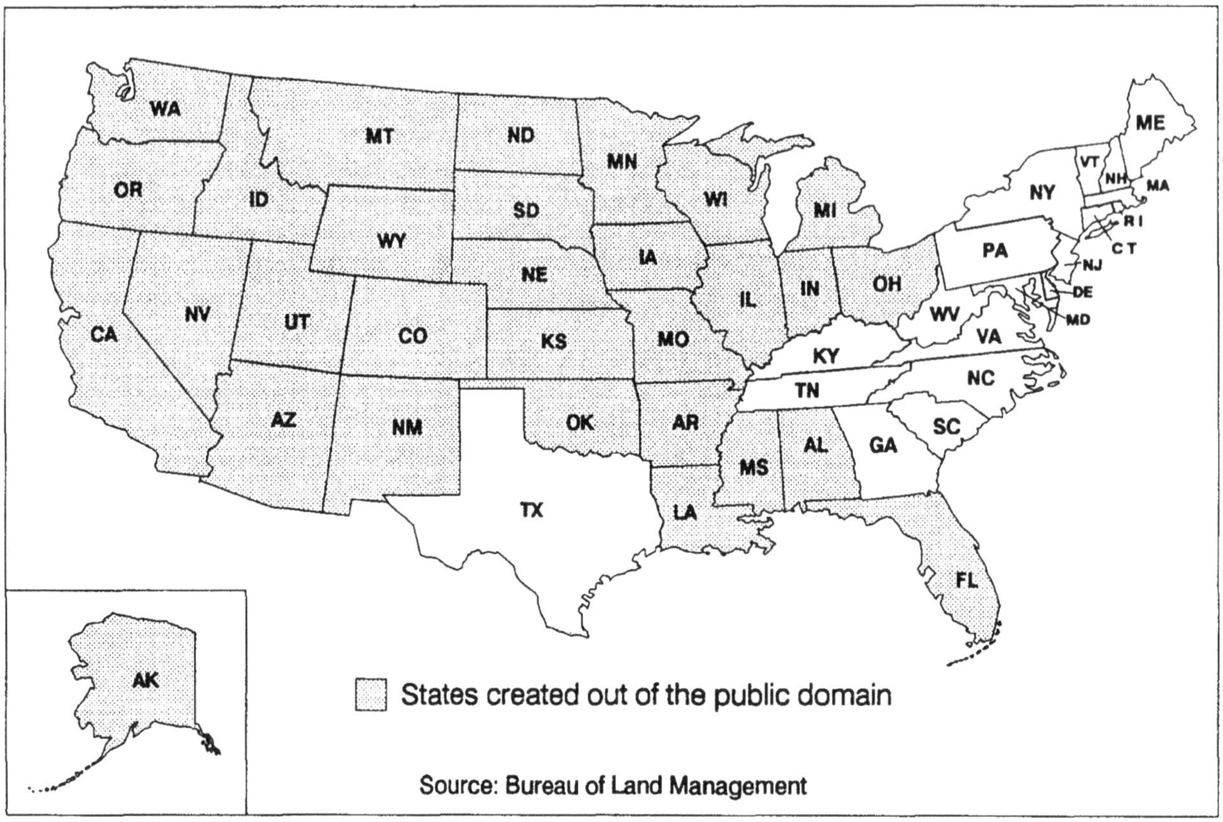

Figure 1-6. Florida was one of the 29 states surveyed from public domain under rectangular method of survey. The original 13 colonies plus Maine, Vermont, West Virginia, Tennessee, Kentucky and Texas retained title to all vacant lands and the U.S. public laws were not applicable.

been given nearly nine million acres of land, slightly more than one-quarter of the total acreage of the State. All of this land was in excess of the 22 million acres given the State by the Federal Acts of 1841 and 1850.

The Federal Act passed by Congress in 1856 stipulated that three railroads must be included in the railroads being proposed. One had to connect the cities of Jacksonville and Pensacola, a second to continue from Pensacola to Montgomery, Alabama and a third to connect the Atlantic Ocean at Amelia Island with the Gulf of Mexico at Tampa. It was realized that without transportation to the interior of the State, the land could not be developed and settlers could not be enticed to settle in Florida.

Soon after the Federal Government passed the Swamp Lands Act of 1850, political leaders in Florida realized that some agency of the State must be responsible for buying, selling, and holding title to state lands. In 1855, the General Assembly adopted An Act to Encourage A Liberal System of Internal Improvements in the State. This legislation was based upon the wording of the First State Constitution that provided it was the duty of the Government of Florida to encourage and provide funds for the development of the State's resources. The Act set up the Internal Improvement Fund, a part of the Cabinet consisting of the Governor, Comptroller, Treasurer, Secretary of Agriculture and the Registrar of State Bonds. The Internal Improvement Fund was empowered to guarantee interest on bonds floated by railroad companies to construct railroads within the State.

In addition to permitting the transfer of title of lands to the railroad companies, the Swamp Lands Act of 1850 also specified that (1) eight sections were granted for a seat of government, (2) section 16 of every township would be granted for the support of public schools, (3) two townships, 72 square miles of land, would be used for seminaries of higher learning and (4) the State would be given five percent of the net proceeds of any land thereafter sold in Florida by the Federal Government.

TOPOGRAPHIC MAPS

In Geography we are interested in space—the manner in which natural and man-made features are arranged within that space. To show how features are distributed over the land and water we use maps. Some maps cover large areas while others cover small areas. The maps covering small areas show more detail than maps covering large areas. One of the most widely used maps used to show detail is the topographic map, commonly referred to as a topo sheet. Topo sheets are published by the United States Geological Survey and include section, township, and range lines as well as latitude and longitude. The maps are drawn in 7½-minute (7½′) and 15-minute (15′) quadrangles and are named for a city, town, or prominent natural feature located within the area. Topo sheets are mainly printed in three colors. The cultural features, such as roads, railroads, cities, and towns as well as lettering are in black; the water features are in blue; and the features of relief, such as hills, mountains, and valleys are shown by brown contour lines. The contour interval (difference in elevation shown by two contour lines) differs according to the scale of the map and the relief of the country. On maps that contain supplemental information, additional colors are used, such as green for woodland areas, and red for highway classification, urban areas, and United States land lines.

An Index Map of Florida shows all areas that have been mapped within the State, the names of the quadrangles, and the date each survey was made. From this Index Map it is easy to find the quadrangle (topo sheet) that covers a particular area.

Florida's oldest topo sheets date back to the 1890's when only seven quadrangles were available: Arredondo, 1890; Dunellon, 1891; Ocala, 1892; Williston, Citra, Tsala Apopka, and Panasoffkee, 1893. Almost two decades elapsed before the Palatka quadrangle was finished in 1912, followed by Interlachen in 1915, and 15 quadrangles around Jacksonville were completed for defense purposes related to World War I in 1917. As late as June 1932, the State of Florida had never shared any of the mapping expenses with the United States Geological Survey and ranked last among the 48 states in contributions. By June 1932, only 24 quadrangles had been completed covering 4,718 square miles of the 58,560 square miles in the State, eight percent of the total area of Florida.

The situation improved by the end of World War II with one-third of the State having been mapped. By 1960 the coverage had increased to 75 percent and finally by 1980 almost the entire State was mapped, the exceptions being in western Nassau and Duval counties and the surface of Lake Okeechobee.

All of the early topo sheets were 15-minute quadrangles. These maps have a scale of 1:62,500 (one inch ≐ approximately one mile on the earth's surface). It requires 296 15-minute quadrangles to cover the entire state. A 15-minute quadrangle, or topo sheet, covers approximately 15 minutes of latitude on its eastern and western boundaries. Along any great circle (a line that completely encircles the earth and divides the earth into two equal hemispheres) one degree equals 69 statute miles.*

Each meridian is a great circle but only one parallel of latitude, the equator, is a

* Statute miles are used to measure ground distance in the United States such as distances shown on road maps. The statute mile contains 5,280 feet. The nautical mile used by planes and ships contains 6,076 (approximately 6,080) feet. Since 1° equals 60 nautical miles (along a great circle), and one minute equals one mile, the nautical mile is much easier to work with than the statute mile. In addition, 60 is divisible by 1, 2, 3, 4, 5, 6, 10, 12, 15, 20, 30, and 60, while 69 is only divisible by 1, 3, 23, and 69.

great circle. All other parallels are small circles. Since 15 minutes is one-quarter of a degree, the distance in statute miles along a great circle is approximately 17.25 miles. Since the only parallel that is a great circle is the equator, any parallel north or south of the equator will be shorter than the 24,902 miles circumference that characterizes the equator. As a result of this one degree will equal less than 69 statute miles. The closer one approaches the poles the smaller will be the value of one degree. At the poles the meridians intersect so there is no distance between meridians thus one degree equals 0 miles. Corrections must be made on topo sheets along parallels to compensate for the changing value of one degree. As a result of the change in value of one degree along a parallel that is a small circle (actually one degree of longitude), the area included in a topo sheet at 20° N latitude would be 17.25 statute miles by 16.25 statute miles and at 30° N latitude would be 17.25 statute miles by 15 statute miles. In each case the area enclosed would be rectangular in shape with the north-south distances longer that the east-west distances.

Most of the topo sheets published since World War 11 have been 7½-minute quadrangles with a scale of 1:24,000 (one inch = 2,000 feet). This means that each side of the quadrangle is 7½ minutes in length, thus it takes four 7½-minute quadrangles to equal one 15-minute quadrangle. However, the same problem of varying lengths of parallels for 7½-minute topo sheets exists as it did for the 15-minute sheets. Thus, a 7½-minute topo sheet exists as it did for the 15-minute sheets. A 7½-minute topo sheet at a latitude of 20° North or South would be 8.625 miles by 8.125 miles, while at 30° North or South Latitude would be 8.625 miles by 7.5 miles.

Topographic maps supply detail usually not shown on county maps or road maps. They are intended to give a picture of the terrain as completely as can be legibly reproduced at the selected publication scale. They can be used as a tool in planning airports, highways, dams, pipelines, industrial plants, and many other types of construction. Topo sheets are essential in flood control, soil conservation, and reforestation. With water supplies so critical in Florida at the present time, hydrologic investigations concerning studies involving quantity and quality of water utilize topo sheets. Individuals also find topographic maps useful in the pursuit of such outdoor activities as hunting, fishing, and backpacking. To appreciate the extent and diversity of the State of Florida, put all of the topo sheets together that have been published.

FLORIDA'S SIZE AND SHAPE

Florida is the second largest state east of the Mississippi River, being smaller than Georgia by only 316 square miles (818 square kilometers). Of the total area 58,560 square miles (152,256 square kilometers), 4,424 (11,449 square kilometers) are water. The state is larger that many nations of the world. For example, it is larger than any country in Central America, as large or larger than Portugal, Austria, Czechoslovakia, Hungary, or Bulgaria in Europe, and is as large as the Netherlands, Belgium, Luxembourg, Switzerland, and Denmark combined. (Figure 1-7).

Shape is also an important characteristic of Florida's physical environment. It is shaped roughly like a boomerang with an east-west distance roughly equal to the north-south distance of 460 miles. However, from Pensacola, near the western boundary, to Key West, the southernmost city, the distance is approximately 830 miles (1,335 kilometers) while the distance from the Georgia state line to Key West is roughly 535 miles (861 kilometers).

The shape of the land above sea level is determined in part by the platform upon which the dry land rests. The platform is know as the Floridan Plateau and consists of both the dry land above sea level and the continental shelf submerged beneath the waves. The boundaries extend seaward to a depth of 50 fathoms (300 feet; 91.4 meters), beyond which the sea bottom drops off steeply. The Plateau stretches southward to the Florida Keys where coral islands flank the steep marine slopes of the Florida

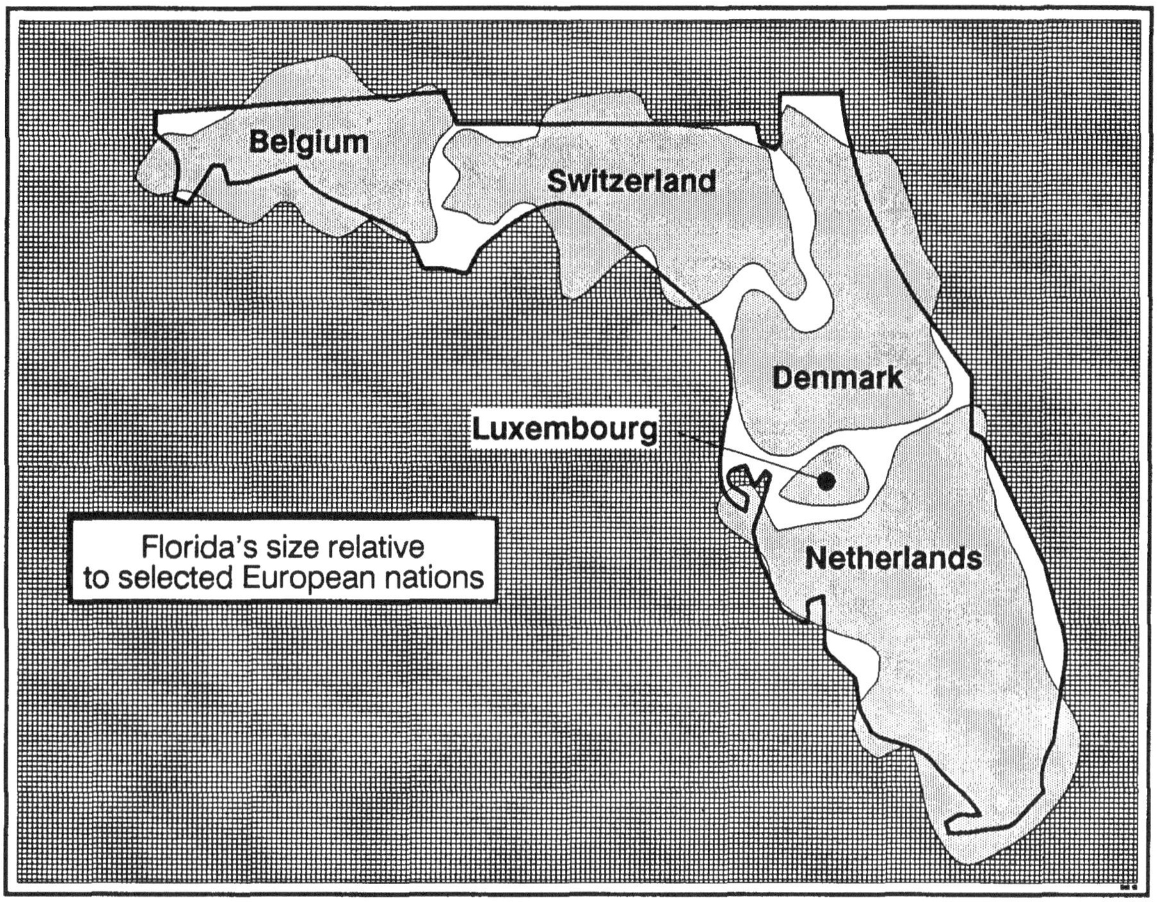

Figure 1-7. Florida is larger than many European countries. The above map shows it is larger than several countries combined.

straits. Cuba lies beyond. Profiles of the Floridan Plateau are shown in Figures 1-8 and 1-9. The platform, shaped like a giant "L" is nearly 500 miles long and 250-400 miles in width. Approximately 58,560 square miles constitutes Florida's dry land and an additional 88,000 square miles submerged land.

The Floridan Plateau originated millions of years ago as part of an island arc that contained an obscure volcanic island in the Gulf of Mexico. This volcanic island became attached to the North American mainland and the union gave birth to the peninsula of Florida. Ocean currents flush with sand deposited their treasures along this new finger thrust in the ancient sea. The finger became a changing, living thing. During the intervening eons nature eroded away the volcanic peaks and leveled the lands. Alternately Florida stood as dry land or was covered by shallow seas. Today the peninsula is recognized as an important platform extension of the North American Coastal Plain. The Floridan Plateau shares a common core of volcanic rock with the Piedmont regions of Georgia and the Carolinas. Florida's modern landscape is an extension of many sedimentary features originating to the north, but much of Florida is unique in all the world.

The boomerang shape attributed to the emerged part of the State consists of part mainland and part peninsula. The mainland part is known as the **panhandle,** the area

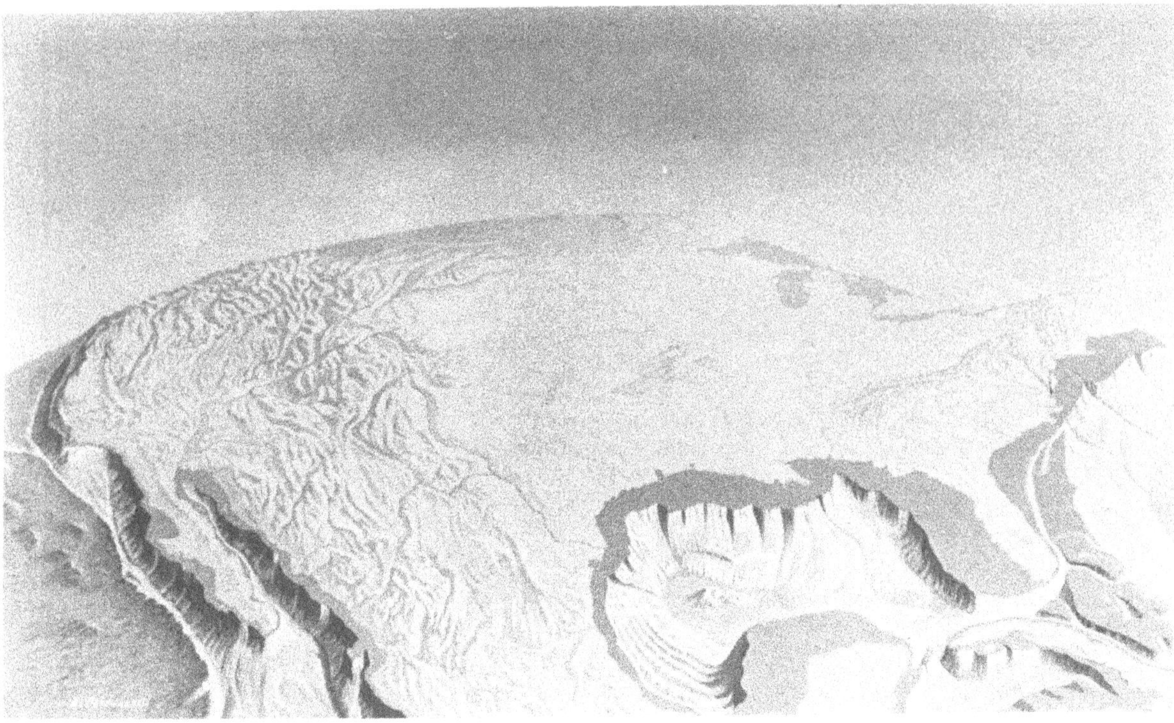

Figure 1-8. Artist's rendering of the Floridan Platform. (Reprinted by permission of Exxon Corporation.)

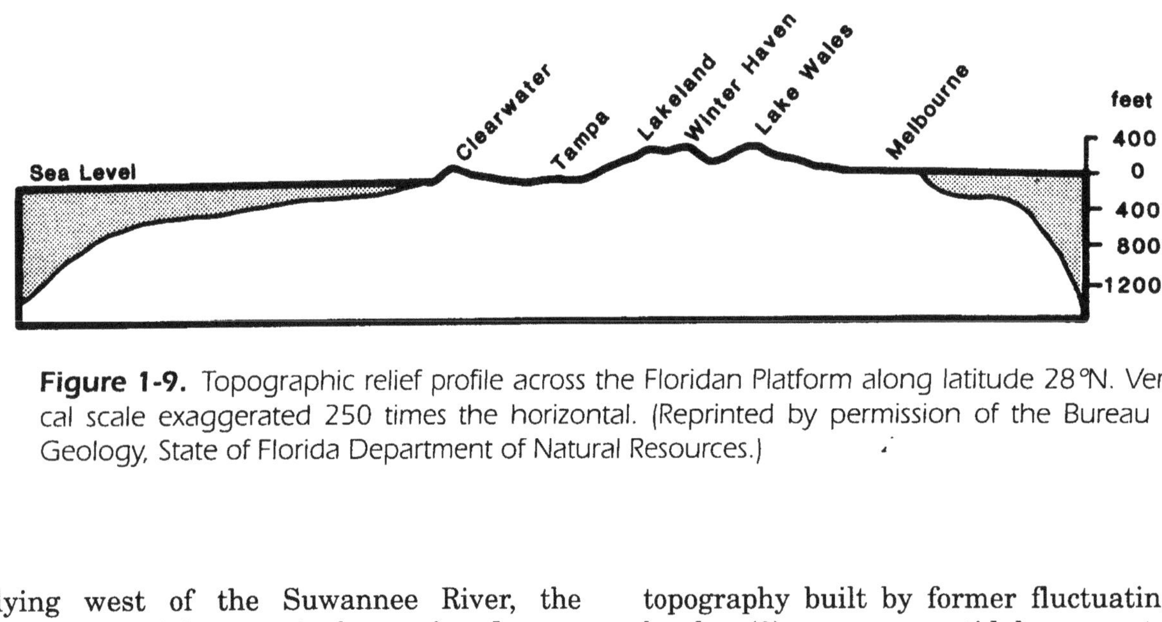

Figure 1-9. Topographic relief profile across the Floridan Platform along latitude 28°N. Vertical scale exaggerated 250 times the horizontal. (Reprinted by permission of the Bureau of Geology, State of Florida Department of Natural Resources.)

lying west of the Suwannee River, the remainder of the state is the **peninsula**.

The peninsula is a nearly flat lowland which gently slopes toward the sea. The highest elevation on the peninsula, near Lake Wales, projects a mere 325 feet above sea level. Many physical features include: (1) a prevailing flat and partially terraced topography built by former fluctuating sea levels; (2) numerous tidal, current, and wind-formed coastal features such as dunes, barrier islands, lagoons, and tidal rivers; (3) abundant coastal and inland lakes and swamps and marshlands and (4) sandy soils heaped into rolling hills and spread out as pine-covered flatland.

The backbone of Florida consists of a line of "sandhills" that extend from the Georgia border to the southern highlands (Highlands County). The flatlands lie on either side of this sandhill ridge and cover a large portion of southern Florida. Coastal features are especially prominent in the south where the tip of the peninsula is highly indented with hundreds of islands spread along the shores.

South Florida contains one of the largest lakes on the continent—Lake Okeechobee. Lake Okeechobee lies adjacent to the Everglades and provides much of the water that gives life to this marshlands overgrown with sawgrass, sedges and tree islands. The combined Everglades-Lake Okeechobee basin is the remains of a shallow sea that existed during the last stages of the glacial epoch.

A key to understanding many of Florida's water-related features lies rooted in the Seminole language. Place names which end in a double "e" are likely to be water related. Okeechobee means "Big Water"; Kissimmee implies "Winding Waters." Characteristically Seminole-named rivers end in "hatchee," thus the Caloosahatchee River means "river of the Caloosa Indians."

The innumerable lakes of the Florida sinkhole country are one of the outstanding peculiarities of the Plateau. Florida's zone of closely spaced, solution basin lakes has no counterpart in North America. The lakes for the most part, occupy sinkholes in which the surface materials have subsided into underground cavities. The lakes, sinkholes and subterranean caverns are all features of Florida's extensive ground water system. In this system, large volumes of water flow through open passageways hundreds of feet below the land's surface. The groundwaters boil to the surface in scattered areas about the plateau. Several discharge points even occur along the seabottom many miles beyond the shore. A typical freshwater spring associated with such underground systems boils to the surface in Tampa Bay.

The Floridan Plateau lies within the latitudinal range containing the world's great deserts. The deserts are a product of dry descending winds associated with equatorial convection cells. Florida uniquely remains unaffected by these desert-producing winds because of its peninsular status between two warm water sources. The result is Florida's warm and humid climate. The mild climate has greatly effected Florida's plant and animal life, resulting in flora and fauna found nowhere else in North America.

The shape and forms of the Florida that we have described appear to be scattered at random about the Plateau. True, there is much diversity here, but as we examine the shaping of Florida under the magnifying glass of time, climate, and nature's forces, a finely sculptured landscape emerges. The scenes of this landscape appear to possess a permanence that is not theirs. Today's face of Florida is here for but a moment in time for Florida is a living and changing land.

REFERENCES

Androit, John L., Editor, (1979), *Township Atlas of the United States*, McLean, VA., Androit Associates, 1184 pp.

Bureau of Land Management, (1947), *Manual of Surveying Instructions*, United States Government Printing Office, Washington, D.C., 613 pp.

Cooke, C.W., (1939), *Scenery of Florida Interpreted by a Geologist,* Florida Geological Survey Bulletin 17.

Dovell, J. E., (1952), *Florida—Historic, Dramatic, Contemporary*, 2 vols., Lewis Historic Publishing Co., Inc., New York.

Marcus, R. B. & Fernald, E. A., (1975), *Florida: A Geographical Approach*, Kendall/Hunt Publishing Co., Dubuque, Iowa, 302 pp.

Marcus, Robert B., (1987), *Study Guide for the Correspondence Course, The Face of Florida*, Department of Independent Study by Correspondence, University of Florida, Gainesville, FL, 269 pp.

Raisz, Erwin, (1964), *Atlas of Florida* University of Florida Press, Gainesville, FL, 52 pp.

Stewart, Lowell 0., (1979), *Public Land Surveys*, Arno Press, New York, 202 pp.

Tebeau, C. W., (1971), *A History of Florida*, University of Miami Press, Coral Gables, 502 pp.

SUMMARY OF CHAPTER 1

Florida is the second largest state east of the Mississippi River and the most southerly of the 48 contiguous states. Despite its latitudinal location in the Desert Belt of the world it is one of the wettest states in the Union because of its position along the southeastern edge of the North American landmass.

The State of Florida rests upon the Floridan Plateau, a platform 400 miles long and 250-400 miles wide that separates the deep waters of the Atlantic Ocean from the Gulf of Mexico. It extends seaward to a depth of 30 fathoms. This platform is millions of years old and originated from a volcanic island that attached itself to the mainland of North America.

At the time Florida entered the Union only 500,000 acres of the state was under the jurisdiction of state government, the remainder was Public Domain. After statehood the land had to be surveyed in order to become available to the public and to the state. Congress decreed that the rectangular system of surveying be utilized in surveying the land. The Base Line and the Tallahassee Meridian cross within the city limits of Tallahassee along Meridian Street. From the point of intersection on the official marker in Tallahassee all "range lines and "townships lines" are calibrated.

Florida has been almost completely mapped by the United States Geological Survey. The maps drawn show the surface of the state and are known as topographic maps or quadrangles. The older maps covered an area of 15 minutes on a side while the newer maps are printed in 7½-minute quadrangles. It takes four 7½-minute sheets to cover the same area as one 15-minute quadrangle.

Name ______________________ Section ____________ Date______________________

REVIEW OF CHAPTER I

Part A

1. Mathematical location can be expressed for a particular area or location by which of the following statements:
   a) Place A is located north of Place B
   b) Place A is located at 29° 11′ N and 81° 02′ W
   c) Place A is located at R33E and T15S
   d) Place A is not located by any of the above statements
   e) Place A is located by statements a & b
   f) Place A is located by statements b & c

2. Relative location places the State of Florida in which one of the following belts?
   a) Tropical belt of the world
   b) Desert belt of the world
   c) Mediterranean belt of the world
   d) None of the above belts

3. Which of the following statements is (are) true?
   a) "Range lines" run in an east-west direction
   b) "Township lines" extend in an east-west direction around the globe
   c) "Township lines" and township have the same meaning
   d) A township consists of an area one mile by one mile and totals 640 acres

4. Just prior to and following statehood for Florida the land that had not been surveyed and still belonged to the Federal Government was known as:
   a) Public Domain
   b) National Forest land
   c) National Seashore land
   d) Swampland

5. Most of the topo sheets published by the United States Geological Survey since World War II have scales of:
   a) 1:650,000 and 1:63,500
   b) 1:62,500 and 1:24,000
   c) 1 inch = approximately 1 mile and 1 inch = 2,000 feet
   d) Selections a & b
   e) Selections b & c
   f) None of the above selections

6. Which one of the following selections is true concerning Florida's area?
   a) Florida is the largest state east of the Mississippi River
   b) Florida is second largest state east of the Mississippi River
   c) Florida has a total area of 58,560 square miles
   d) Selections b & c are true
   e) None of the above selections are true

Part B

1. What is meant by?
   a. Mathematical location?

   b. Relative location?

2. a. From Goode's *World Atlas*, a local atlas, a road map or some other source, determine the latitude and longitude (within ½°) of the town in which the high school from which you graduated is located. If the high school was a consolidated high school in a rural area, use the town nearest to the high school to answer this question. Give the name of the town and the high school.

   b. What is the Range and Township of the following cities?: Panama City, Tallahassee, Gainesville, Daytona Beach, Fort Myers, and Palm Beach.

   c. Give the relative location of the town in which you are now living. (Hint: use four adjacent towns, one for each of the four cardinal directions in describing the relative location of your home town)

3. a. What is a topographic map?

   b. How many times larger is a 15-minute topographic map than a 7½-minute topographic map?

4. a. Differentiate between the term range and "range lines."

b. Differentiate between the term township and "township lines."

5. For each of the following exercises, state your source of data and give the area to the nearest 100 square miles. Use the area for Florida and for each of the states or countries in making your comparison.

a. Compare the area of Florida with the area of California.

b. Compare the area of Florida with the area of Texas.

c. Compare the area of Florida with the area of the Netherlands.

CHAPTER 2
Paleogeography of Florida

INTRODUCTION

The face of Florida is not static, but dynamic. It is continually changing. The natural forces of running water are wearing down and eroding its surface. Sinkholes open to create new lakes and ponds. Earth stresses cause the land to slowly buckle, rise, and fall. Blowing sea winds build dunes along the coasts, while changing ocean currents remove the beach sands. Man, with his machines and his activities, is altering the Florida landscape too; often at a rate exceeding the processes of nature.

Through the eons, Florida has alternately stood high above the waves and submerged below primeval seas. Florida's form and shape once was radically different than its face of today. And what will be the face of Florida tomorrow? Satellite measurements tell us that sea level is currently rising at the rate of nearly one-quarter inch a year as the earth's polar ice caps continue to melt. The impact of this process alone will reflect on the shape of Florida's land forms in just a few short years.

The study of geography of ancient times is termed **paleogeography**. The following study of Florida's paleogeography concerns the various land forms, seas, climates, and life forms that comprised the state from hundreds of millions of years ago to the near-present.

Earth history is divided into a series of geological time divisions based on the magnitude or impact of certain natural events. The largest time division is termed an era and is recognized as representative of a major chapter in earth history. Individual eras are separated by major events that affect the lands and life of the whole planet. For example, at least three of the four eras are separated by earth's collisions with giant meteorites, which smashed continents and caused mass extinction of the then major life forms. Each era is divided into smaller divisions called **periods** which are separated by minor mountain building and localized geologic events. Examples of such division events include the uplift of the Acadian Mountains of the American Northeast and the flooding of the middle of North America.

Table 2-1 reflects the various divisions of the Geologic Record. Continental studies have shown that the divisions of the Geologic Record cannot be readily based on widespread geologic changes, unconformities, etc., as originally believed. However, the Geologic Record has been retained for chronological description purposes.

Table 2-1. Geologic History of Florida

ERA	Period	Epoch	Years Ago	Major Geological Events in Florida	Main Lifeform of ERA
Precambrian			2,000,000,000	Oldest rock in Florida found thousands of feet beneath the surface.	
Paleozoic			590, 000,000		
	Cambrian		500,000,000	Northern half of Florida probably stood continuously above a fluctuating seaway whose depth and extremities fluctuated noticeably. Total emergence may have existed for short geologic time intervals. The stability of the Floridan Platform was promoted by its volcanic activity and by the formation of the Peninsular Arch. Fossil studies conducted for this era indicate climates during these years were mostly warm and humid.	**Marine Invertebrates**
	Ordovician		427,000,000		
	Silurian		400,000,000		**Fish**
	Devonian		345,000,000		
	Mississippian		310,000,000		
	Pennsylvanian		280,000,000		**Amphibians**
	Permian		230,000,000		
	Triassic			With emergence of the Appalachian low lands into a mountainous belt, Florida was contemporaneously connected to the main portion of the continent. All of the Triassic state (with exception perhaps of the Keys) stood firmly above sea level for most of the period and experienced a temperate climate. Land surfaces eroded significantly.	
Mesozoic			188,00,000		**Reptiles**
	Jurassic			Climates moderated and seas advanced on a worldwide scale. Florida was inundated.	
			135,000,000		**Fish**
	Cretaceous			Complete submergence of Florida to moderate depths. Sediments in the northern portion are shallow water types, but become progressively more marine southward. This indicates southern Florida lay more deeply beneath the seas.	

ERA	Period	Epoch	Years Ago	Major Geological Events in Florida	Main Lifeform of ERA
			63,000,000		
		Paleocene			
			58,000,000		
		Eocene		Early in the period, seas advanced further northward beyond Cretaceous limits. Retreat of the seas began during the middle of the Tertiary. Portions of Florida may have emerged above sea level along the Ocala uplift which formed during this period.	
			36,000,000		
		Oligocene			**Mammals**
Cenozoic	**Tertiary**		25,000,000		
		Miocene			
		Pliocene			
			1,000,000		
		Pleistocene		Portions of the state were alternately exposed by glaciation-induced sea level changes. Terrace remnants mark old shorelines.	**Man**
	Quaternary		10,000		
		Recent			

FLORIDA AND PLATE TECTONIC CONCEPTS

The geomorphic or land form histories of Florida and southeastern North America are closely intertwined, and perhaps may be better understood in a regional context. No one was present to record most of these histories, and thus they must be carefully compiled from the bits and pieces of assorted geologic remnants. Ancient outcroppings are common in the Appalachian Mountains and adjoining Piedmont areas, but the geologic record of the coastal plain lies buried beneath thousands of feet of sediment. These buried records can be reached only by deep drilling, especially in Florida.

The regional Paleozoic history (from 600 million to 230 million years ago) that is presented here is based upon the latest geologic theories—much of which has only moderate substantiation. These theories are based upon Hatcher's (1972) Appalachian model, the concept of plate tectonics, and recent evidence of meteorite collision with earth. The reader should note that the evaluation of the surface forms described has not followed a consistent path, but has proceeded in a series of pulses, retrogradations, and minor erratic changes. However, the net progression is basically as presented.

For those unfamiliar with the concept of plate tectonics, the crust or rigid surface of the earth is broken into a series of tectonic plates as shown in Figure 2-1. For the sake of simplicity, the plates may be visualized as floating on the **asthenosphere** or subsurface zone of heated, plastic material and liquid rock (called **magma**) and propelled about slowly by convection currents. With time, moving plates may drift apart or collide with one another. Diverging plates open fissures along the separation line. Surface-flowing molten rock (termed **lava**) may then flow from these fissures as shown in Figure 2-2. When plates collide, the edges of the plates may crumple and fold to form mountains in a process called **suturing** (Figure 2-3). The Himalaya Mountains were formed in this manner. An alternate result of collision may result in one plate being deflected downward beneath the other in what is termed **subduction**. In the zone of subduction a deep trench forms forward of the actual convergence of plate boundaries (Figure 2-4). The Japan and Marianas Trenches exemplify subduction zones. Two major types of rocks associated with this type of activity are **plutonic**, which cool from magma, and volcanic, which solidify from cooled lava.

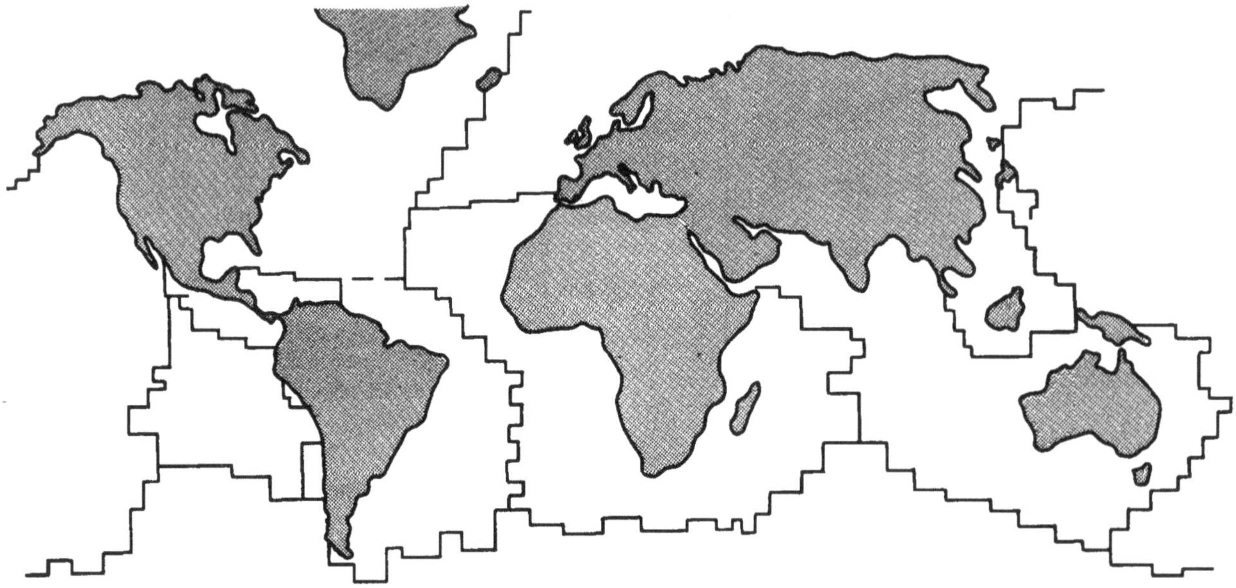

Figure 2-1. Major tectonic plates as they appear today. Smaller plates are not shown. The boundary between South American and North American plates is uncertain.

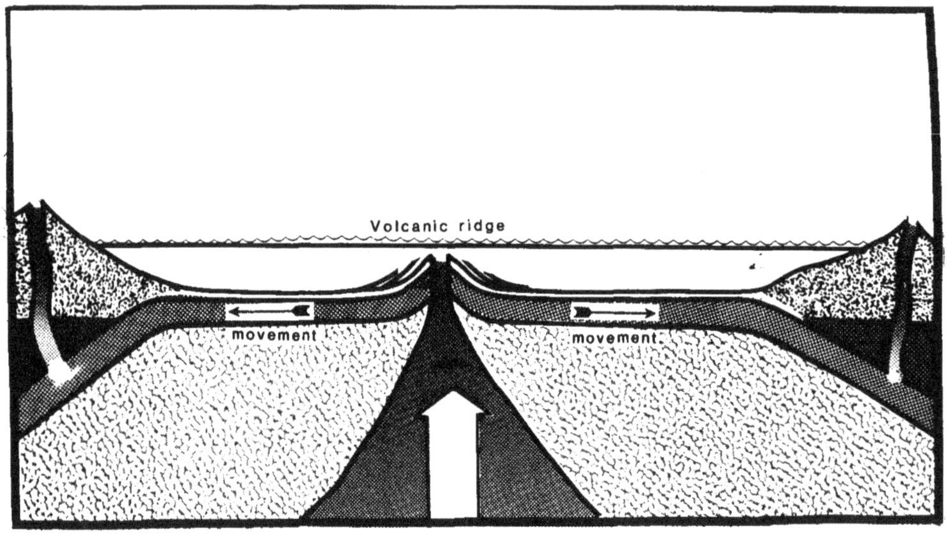

Figure 2-2. Plate divergence produced a series of tensional rifts in the oceanic crust. Lava flowing from these rifts created complex ridges on the sea floor. The lava-covered ridges were repeatedly fractured by transform faults, resulting in offset ridges.

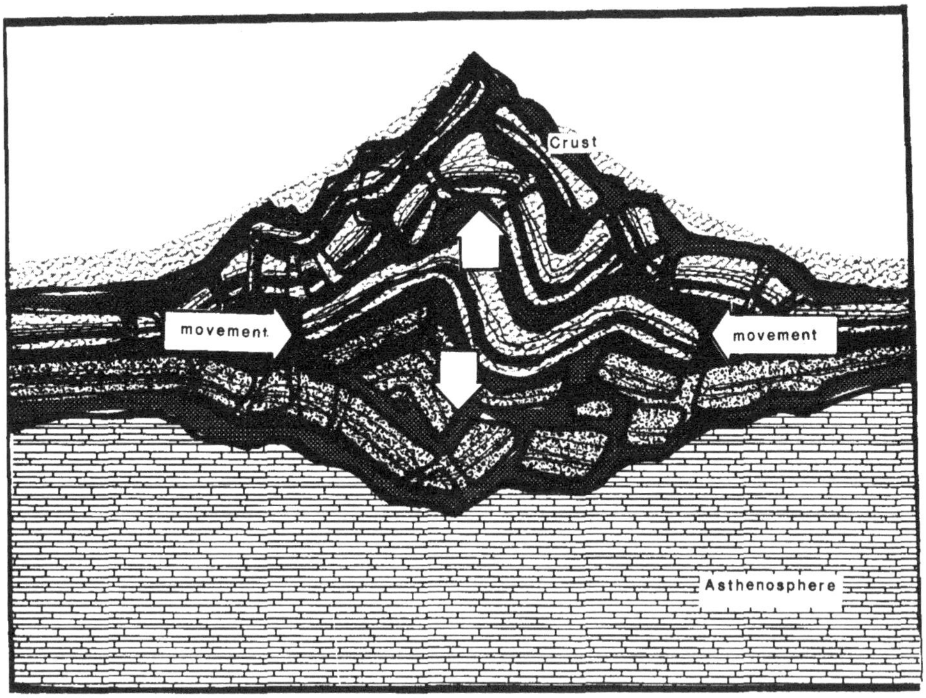

Figure 2-3. Folded and faulted mountain belts result from the collision of two continental platforms.

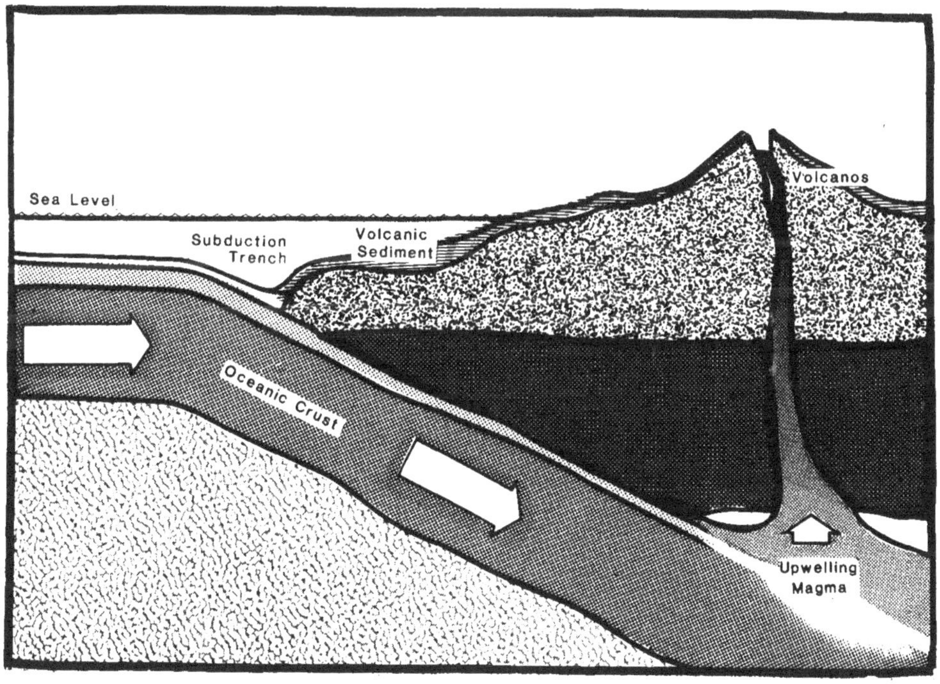

Figure 2.4. Subduction of an oceanic plate beneath Appalachia. Low-density magma produced by melting along the subduction zone rises buoyantly to form a series of volcanic mountains and ridges.

PALEOZOIC TECTONICS AND PALEOGEOGRAPHY

The North American Plate of the Paleozoic Era was drastically different from the way that it appears today. North America was fragmented into a series of islands that alternately joined and separated with the rise and fall of both sea level and dry land surfaces. Florida lay at a bend in what is called "the Continent of Appalachia." Appalachia stood persistently throughout the era as an elongated plateau extending from Canada south to today's Texas coast. On a continental scale, Paleozoic land changes altered Appalachian boundaries only to a moderate extent. During portions of that era Appalachia was connected to Canada and Mexico. Appalachian borders shifted west as Precambrian mountains to the east eroded and mountain building raised former geosynclines or extensive undersea troughlike depressions. However, the general configuration of the continent per se was not significantly altered.

During the early Paleozoic, the European, African, South American, and North American tectonic plates are believed to have drifted toward one another at a rate of 2-10 inches a year. The African plate struck North America along what is now the eastern seaboard of the United States. Europe collided with the Canadian Maritime Provinces and New England. As a point of reference, New York and Spain were joined; their combined lands lay along the equator. Florida, West Africa, and Brazil were sutured together in an area approximately 10° south of the equator. Geologists have named the massive continent formed by these multiple collisions **Pangaea**. It has also been suggested that before the plates collided, a micro-continent which they called Avalon existed in the "Proto-Atlantic Ocean" between the North American and African continents (Figure 2-5).

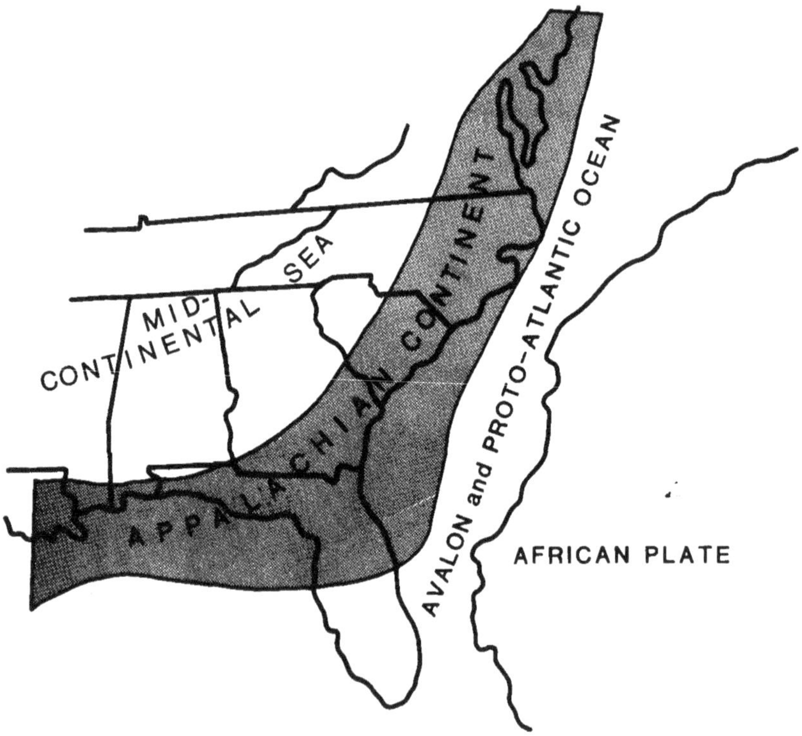

Figure 2-5. Closing the Proto-Atlantic Ocean and collision of the African and North American plates. The land of Avalon is believed to have existed between the two continental platforms prior to the collision. Data on the profile of Africa during this period are lacking.

It is believed that the African Plate, or Avalon, deflected or bent downward on entering the subduction zone. The deflection or bending zone was about 60 miles wide and extended from Labrador well into the Caribbean, a distance of some 1,600 miles along the present United States coast alone. The affected area along the United States eastern seaboard is called the Appalachian continent. The deflecting plate then slid into the plastic zone of the asthenosphere at a steep angle. Friction and subsurface heat caused the subducting edge to melt upon reaching a depth of 100 to 200 miles. The resulting molten rock (or new magma) rose toward the surface due to density differences in a manner much like oil on water. This molten material eventually broke through the overlying crust of the American plate. The result was a series of volcanic island arcs extending along the Appalachian coast and south at least to the Old Antilles and Cuba. Similar subduction features are believed to have occurred beneath Louisiana and eastern Mexico due to the collision of the two American plates. The pattern of island arcs and related features is shown in Figure 2-6.

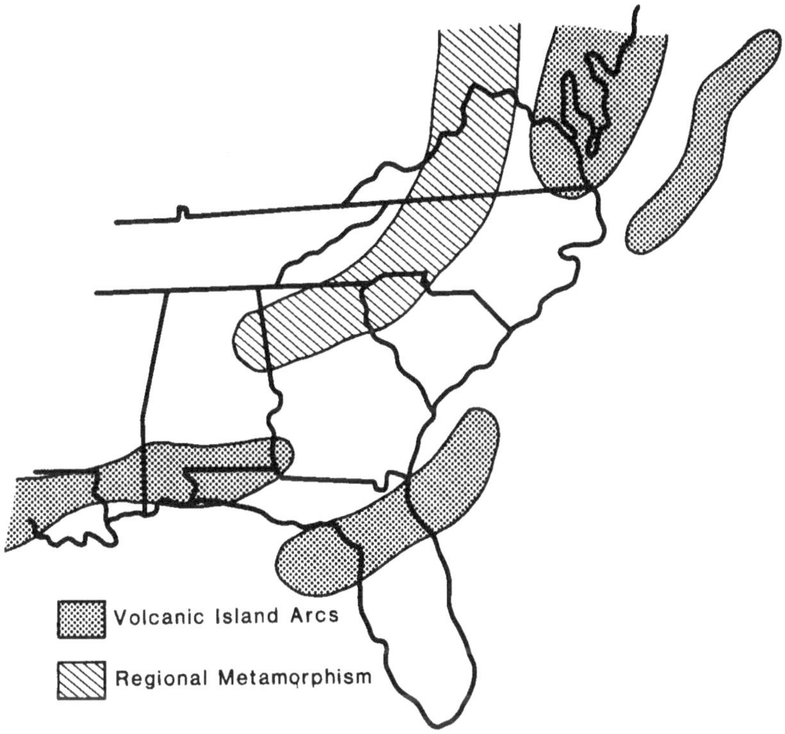

Figure 2-6. Volcanic mountains formed along the edges of the Appalachian continent and as off-shore island arcs during the Paleozoic period. The volcanism accompanied the subduction processes. Regional metamorphism also occurred during this era as shown.

Characteristically, volcanos that are associated with island arcs and subduction zones occur as vent-type volcanos that erupt explosively and produce large lava flows. Rock remnants suggest Appalachian volcanic islands were of the explosive type.

Studies of Florida's subsurface geology indicate the Floridan Plateau was similarly affected by the collision. It is believed a series of island arcs formed in central Florida. Remnants of these arcs have been located below Lake, Osceola and Volusia Counties. These remnants include granite, rhyolite, and tuff (types of plutonic rocks which cooled from magma beneath the crust). The volcanic islands, which formed above this plutonic material, were separated from the Appalachian mainland by what is called the "North Florida Trench." Drilling suggests the Trench was approximately 150 miles wide and deepest close to the continental margin.

Large masses of molten magma associated with the irregular volcanic activity produced repeated island uplifts through the early middle Paleozoic Era. Seismic events later in the era resulted primarily in magmas trapped beneath the surface. Heat accompanying intruding magmas altered the chemistry and texture of the surrounding rocks. This process is called **metamorphism**. Drilling indicates some metamorphic rock remains in conjunction with Florida's basement rocks.

Florida's older metamorphic events probably occurred 1,050 to 1,309 million years ago. Florida's basement rocks are essentially the same as those found in the Piedmont and Blue Ridge Areas to the north.

Florida's Paleozoic island system was probably comparable to the Japanese islands of today. Japan is representative of a modern subduction-related volcanic system. Like Japan, Florida's volcanic mountains undoubtedly pushed upward hundreds of feet above the surrounding seas. The islands were composed of ash, lavas, and volcanic debris. Natural weathering probably produced rich soils similar to those found along the flanks of old volcanos in our western states.

The geologic record indicates a definite lack of climatic zones during the Paleozoic Era. The Floridian islands, like much of the world were marked by a relatively warm and humid climate. This was due to the moderating effect of widespread shallow seas. Life in the polar seas was quite similar to that of much lower latitudes. Corals similar to those along the Florida coasts were also common in Labrador waters as well. The lack of growth rings in woody plants of the era suggests little seasonal variations. Near the close of the Paleozoic Era, especially during the Permian Period, aridity and alternative wet-dry climates became more prevalent throughout North America. However, Florida probably maintained its humid climate throughout the age due to its warm surrounding seas.

The fossil record of Paleozoic life in Florida has been almost completely removed by erosion. However, it is presumed identical to that of the adjacent Appalachia, where fossil remnants do remain. Plant life was prolific throughout the Paleozoic Era (and for eons before) although few actual plant remains have been recognized.

The oldest land plants have been identified from the records of 400 million years ago. During that period, they ranged from small herbs to swamp and moisture-loving trees growing over 40 feet high. The abundance of vegetation remains from this age forward suggest that this was the first geologic time period when much of the land (presumably including Florida) was blanketed by forests. Paleozoic forests were comprised of large scaled trees up to three feet in diameter. The forests also contained small to large ferns (some over 50 feet high); horsetails—a plant similar to bamboo but often reaching over 100 feet in height; and short-needled, pine tree-like conifers.

A variety of mobile life forms lived and fed among the plants of this age. Scorpions, spiders, and a diversity of insects scurried, leaped or flew among the plants. Many of the insects were much larger than those of today. Four-inch cockroaches and dragonflies with 30-inch wingspans were common.

The first known land vertebrates appeared in the middle to late Paleozoic times. Amphibians probably crawled ashore on the Florida islands during the Devonian Period. The fossil record indicates the smaller ones were only one to four inches long, while longer ones reached alligator-size. Considerable variety existed among these creatures. The slower ones waddled on stubby legs while the more lizard-like were quick and active. More marine forms possessed snake-like bodies. All were carnivorous. They fed on insects, fish and one another.

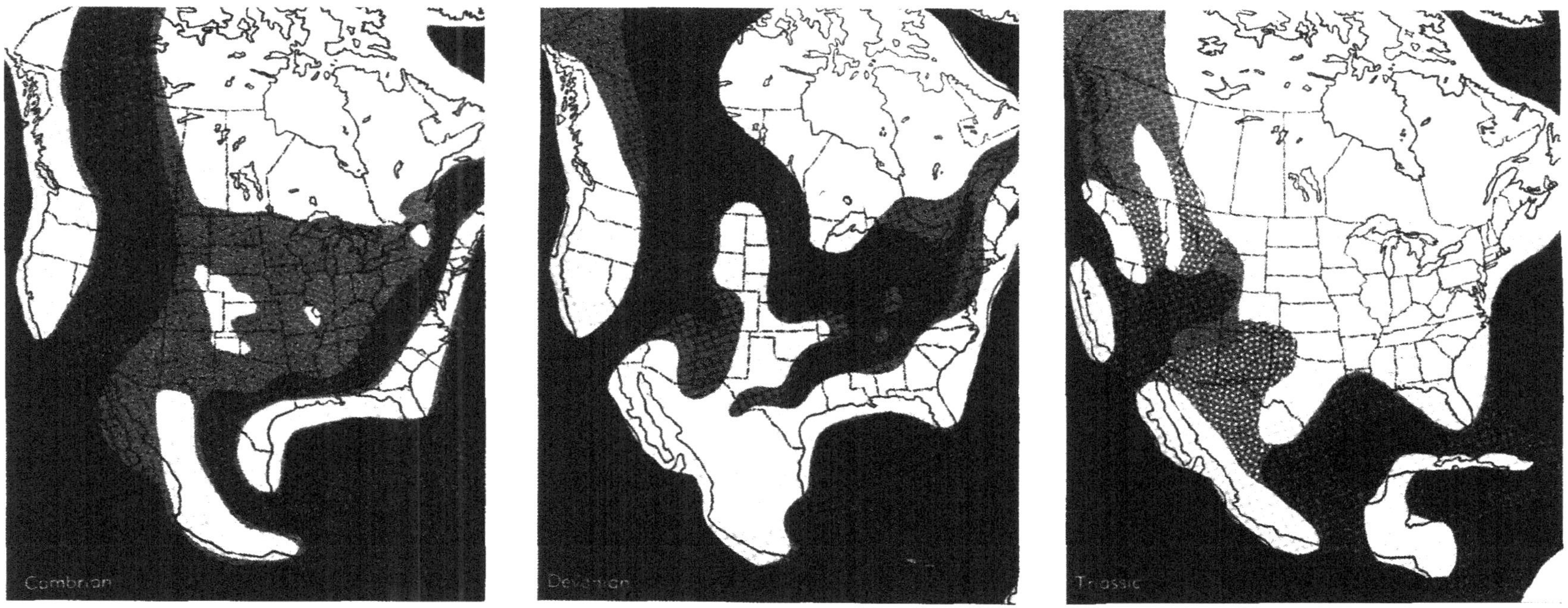

Figure 2-7. North American continent during 3 periods of the Paleozoic Era. Persistent dry lands are shown white; oceanic areas in black; more persistent seaways on the continental platform are indicated by dark grey; and areas submerged during only part of the period are represented by light grey. Areas of temporary marine flooding during the Triassic are dotted light grey. Remember, each of the periods shown lasted millions of years, so such changes over these long time frames are to be expected. (Source: Moore, Introduction to Historical Geology, McGraw-Hill Companies, 1949.)

Figure 2-8A. Flowerlike crinoids catch small sealife while a straight-shelled cephalopod devours a small trilobite, as a larger one crawls by. (Buffalo Museum of Science, Neg. 23658)

Land dwelling reptiles and perhaps early mammals appeared in increasing numbers toward the end of the period. The reptiles emulated the amphibians in diversity of forms. Many of the agile varieties were carnivorous, with pointed teeth for catching and tearing. The slower moving "waddlers" tended to grub on leaves, roots, and stems and an occasional fish or other small invertebrate.

The aquatic life in Florida's fresh and marine waters was even more diverse. Shellfish, sponges, and corals abounded in the shallow seas. The oldest known fossil-bearing beds contain trilobites in such numbers that they must have been the most abundant organisms of the Cambrian age. These creatures are probably the predecessors of today's horseshoe crabs. The number of trilobite varieties has declined through the ages,

Figure 2-8B. Terrestrial and aquatic plants of the Devonian period. The horsetails shown grew abundantly in fresh ponds. The tallest plants are tree ferns which reached a height of nearly 50 feet. The shorter ones are leafless adaptations of algae which grew in moist soils to heights of 2-3 feet. (The Field Museum, #73898, Chicago.)

Figure 2-8C. Skeleton and restoration of a Carnivorous Paleozoic reptile with a fin back. Some of these attained a length of 8 feet or more. These land-dwelling reptiles became extinct before the end of the Permian period. (Top: Neg. no. 315862. Courtesy Dept. of Library Services, American Museum of Natural History. Bottom: Neg. No. 35741. Courtesy Dept. of Library Services, American Museum of Natural History.)

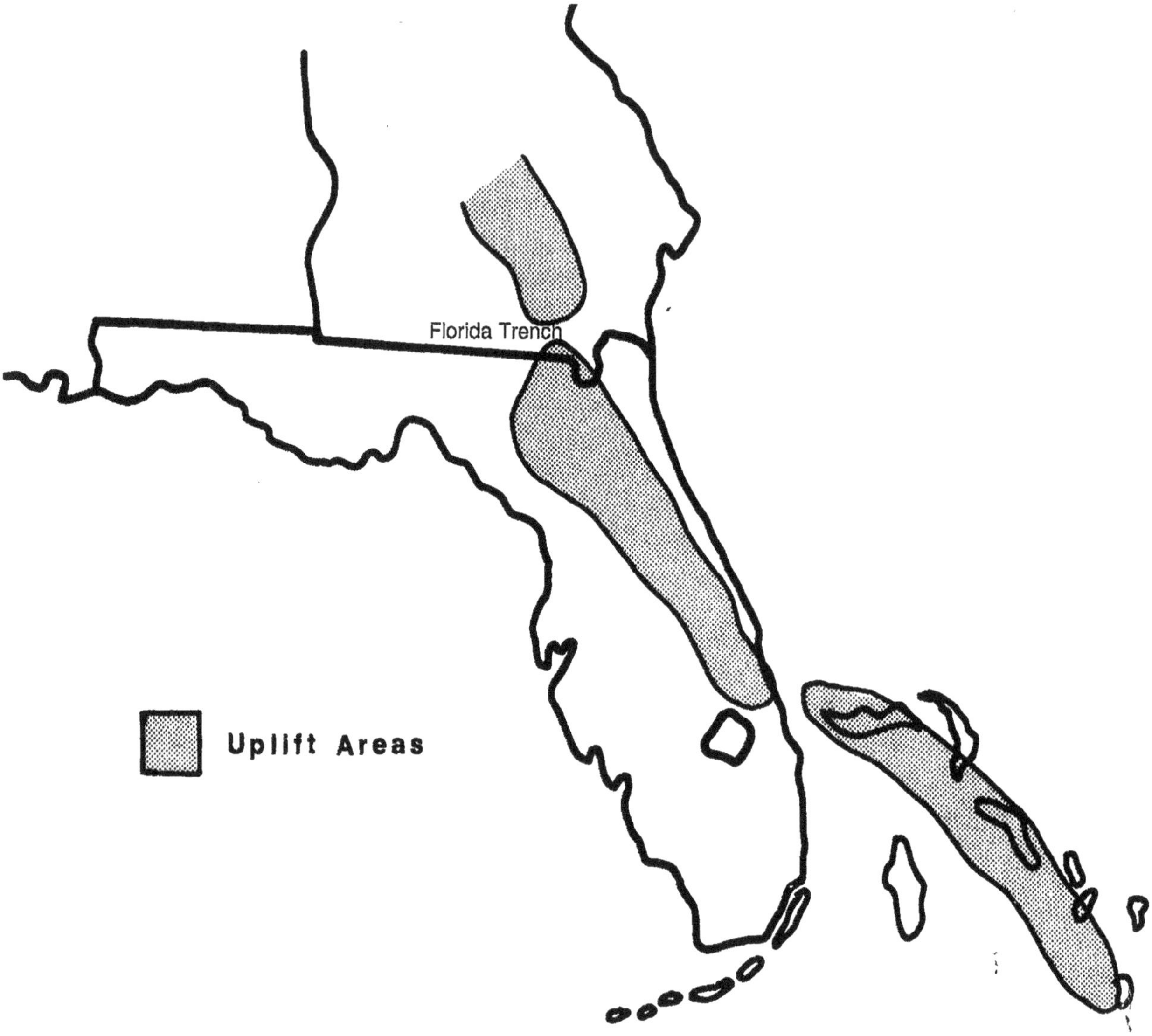

Figure 2-9. Florida's Peninsular Arch is possibly part of the Appalachian folded structure containing the Central Georgia Uplift and the Bahama Islands.

from well over 1,000 in early times until their disappearance in the Permian climax of Fish undoubtedly ruled the seas surrounding the volcanic islands. The fish became prolific in fossil records of the Devonian Period, midway through the era. None of the strata of this era contains modem sharks or rays, but evidence of their forerunners has been recognized.

During various stages of the Paleozoic Era, periods of uplift created new mountain belts throughout Appalachia. The uplifts are explained in part by the continued collisions and rebounding of the plates described. The new ranges were formed by the uplifting of former erosion-lowered mountains, plains, and marine sediments. These repeated uplifts maintained a mountain axial chain extending the length of Appalachia. Erosional sediments from Appalachia continued to accumulate along the coastal plains and the Florida islands. These land based sediments are termed **clastic** sediments, and over the ages filled much of the "North Florida Trench" between the islands and mainland Appalachia.

Florida's Peninsular Arch (see Figure 2-9) represents a Paleozoic anticline or upward folded structure. It is this Peninsular Arch which forms the axis of today's Floridan Plateau.

Toward the end of the Permian period Pangaea began to pull apart. North America pulled toward the northwest, while the African-South American land mass pulled toward the South splitting into two continents. The breakup of Pangaea undoubtedly did not occur along the actual subduction zone that bound the supercontinent together. No actual trace of the subduction zone remains. Watkins (1970, 1971) suggests the zone follows the Brevard Fault Zone extending from Georgia through Virginia. No trace of the Brevard Fault Zone is recognized in Florida. Other investigators suggest the subduction zone currently lies buried beneath Atlantic sediments.

Some investigators believe that as the Atlantic Ocean opened, portions of Avalon, Africa, and South America remained welded to the North American Continent. Geologists apply the term **terranes** to such continental fragments. The basement rock of Florida suggests the southern peninsula is actually a terrane pulled from South America. North Florida, South Georgia-Alabama appear to be pulled from Africa. Evidence suggests that portions of the North American plate in the area of Mississippi and east Texas were stolen away to form Mexico's Yucatan peninsula. The remaining hole became the Mississippi valley. The terrane concepts described are based upon fossil similarities and the alignment of magnetic particles frozen in the rocks. The Florida terranes are believed to consist of volcanic rocks overlain with sediments containing fossils of lower Ordovician to middle Devonian Age.

It is difficult to determine when and why Pangaea began to break apart. This is because during the last 250 million years, many of the clues have been buried or removed by normal (and abnormal) geologic processes. Geologic history does suggest that Pangaea broke originally into two super continents of Laurasia and Gondwanaland; and these in turn divided into small continents similar to those of today.

Geologists currently argue as to whether the breakups and subsequent planetary changes resulted from volcanic activity, collisions of asteroids, meteors, and comets, or a combination of these. The combination theory is perhaps the most accepted.

The latest scientific theories suggest an asteroid may have struck Siberia, creating massive flood-like eruptions of fiery lava. The site is near Lake Baikal and is called the "Siberian Traps." The site is some 870 miles in diameter, and the lava flowed for nearly 600,000 years. These Siberian eruptions were probably extremely violent, blasting large amounts of dust ash, and sulfur gas into the air, where the jet stream carried them around the earth. The sulfur gases undoubtedly mixed with moisture in the air returning both to the earth in prolonged acid rains. The ash and dust reduced the available solar radiation and created ice ages.

Additional rock evidence found by NASA indicates another asteroid, 6 to 12 miles in width, cracked Gondwanaland, and promoted its breakup into South America, Antarctica, Australia, India, and South America.

Each of these catastrophes resulted in massive extinction's of wildlife. The fossil record indicates more than 75% of the marine species were lost, including all the trilobites and many corals, shellfish, and fish types. On land many of the plants, animals, and amphibians disappeared. However this "Great Dying" opened the planet to the appearance of new species, including the dinosaurs of the following Mesozoic era.

PALEOGEOGRAPHY OF THE MESOZOIC AGE

The Permian decoupling of Africa and North America continued to influence the respective plate geologies. Both plates expanded their boundaries as sea floor spreading attached new materials to their trailing edges. Cooling and shrinkage of new lithospheric surfaces proceeded gradually as the plates diverged or separated.

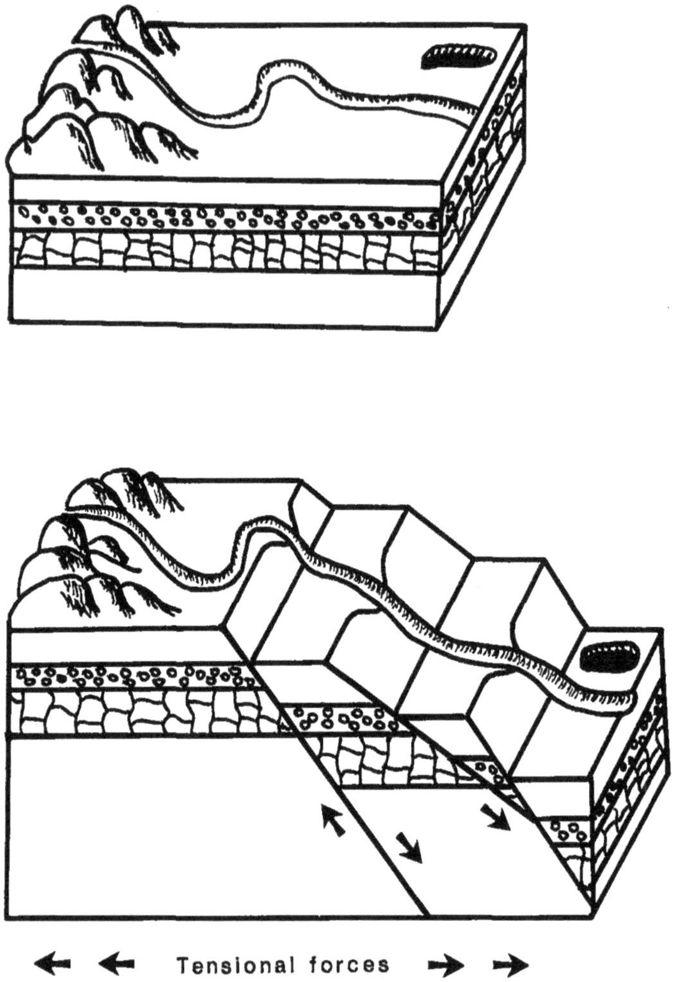

Figure 2-10. Top: Broad reference plane undeformed by tectonic activity. Drainage begins in the mountains to the left; flow is toward the right. Bottom: Normal faults produced by tensional forces extend the topographic surface and roughen the character of the landscape.

Plate separation created tensional stresses within the rock of both American and African plates. The stresses were expectably greater in areas closer to the zone of divergence. Tensional stresses accompanying divergence zones typically produce normal faults. Such faulting occurs along inclined fracture planes as shown in Figure 2-10. The result is a lengthening or extension of the plate surface and a roughening of the topography.

Evidence suggests that Mesozoic faulting and folding produced a series of far-reaching changes in the form and landscape of Florida. Plate divergence probably caused tension fault activity on all sides of Florida's plutonic basement. Faulted steep marine slopes appeared along America's east and Gulf coasts as Pangaea broke apart.

Faulting and folding along the Plateau's northern front created two deep coastal embayments of Figure 2-11 the Okefenokee embayment in the Atlantic basin and the Apalachicola embayment of the Gulf basin. The Apalachicola embayment covered an area of approximately 30,000 square miles. The Okefenokee embayment was smaller and more shallow. An erosional channel

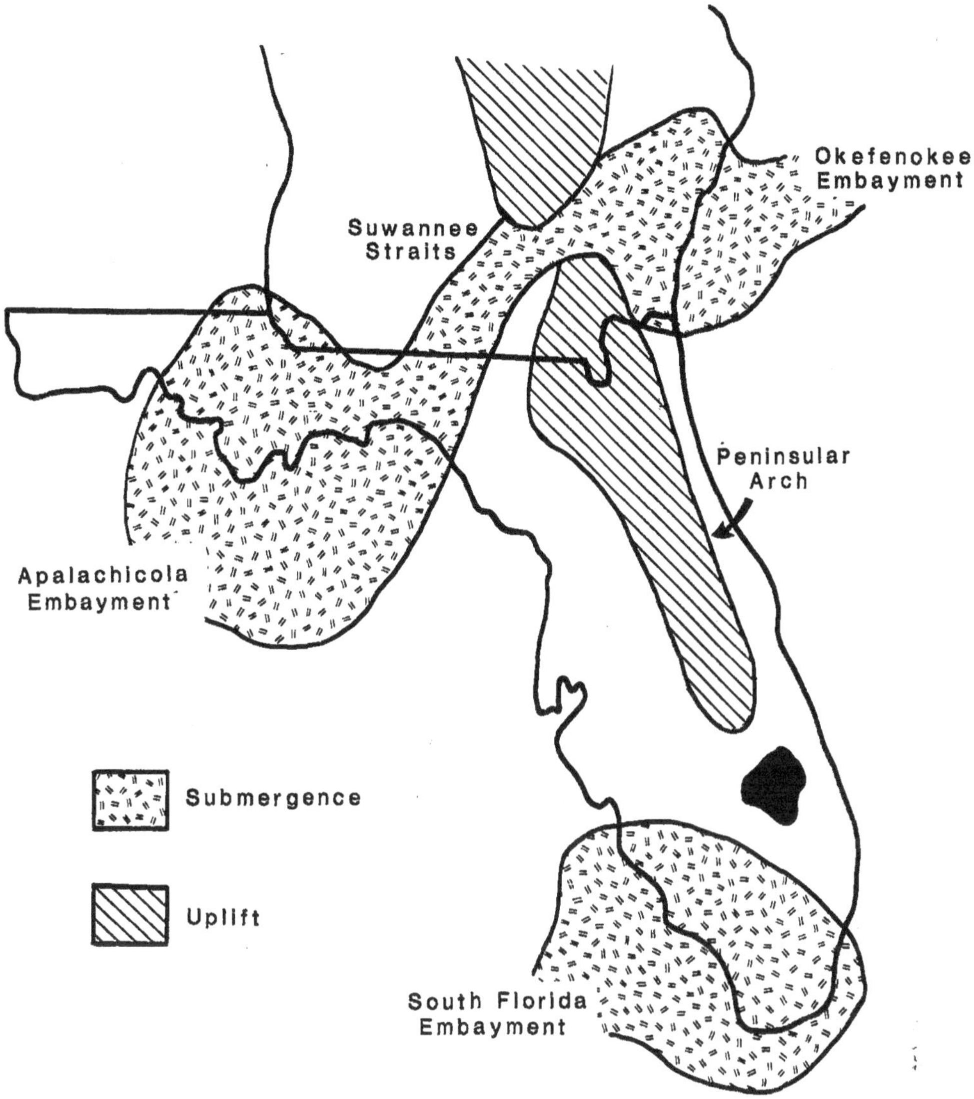

Figure 2-11. Major geomorphic features of Florida during the Mesozoic and early Cenozoic Eras.

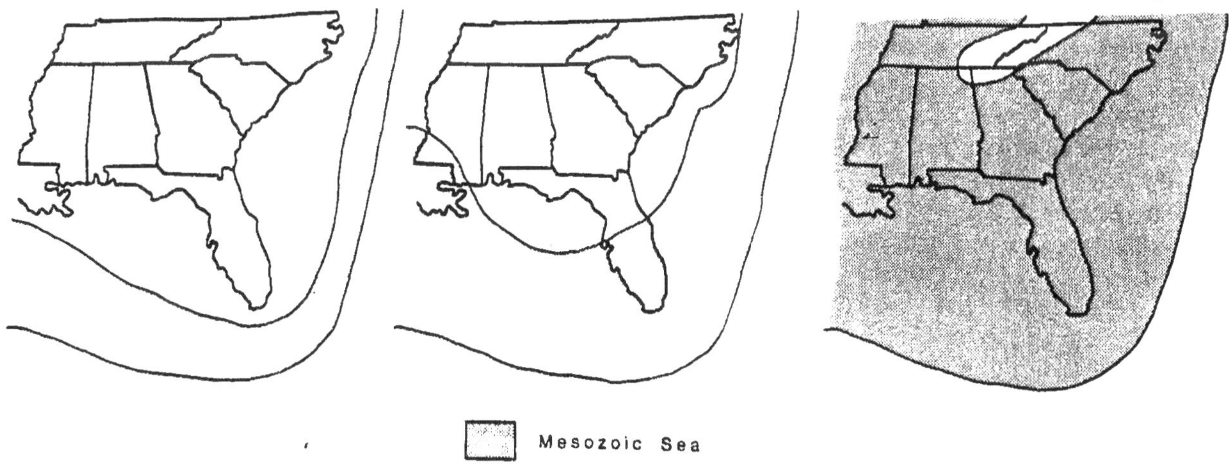

Figure 2-11A. Contemporary sea levels during various periods of the Mesozoic Era. The seas fluctuated during these periods, but the overall trends toward inundation of the North American continent proceeded from the early stage (left) to the final cretaceous ocean covering on the right.

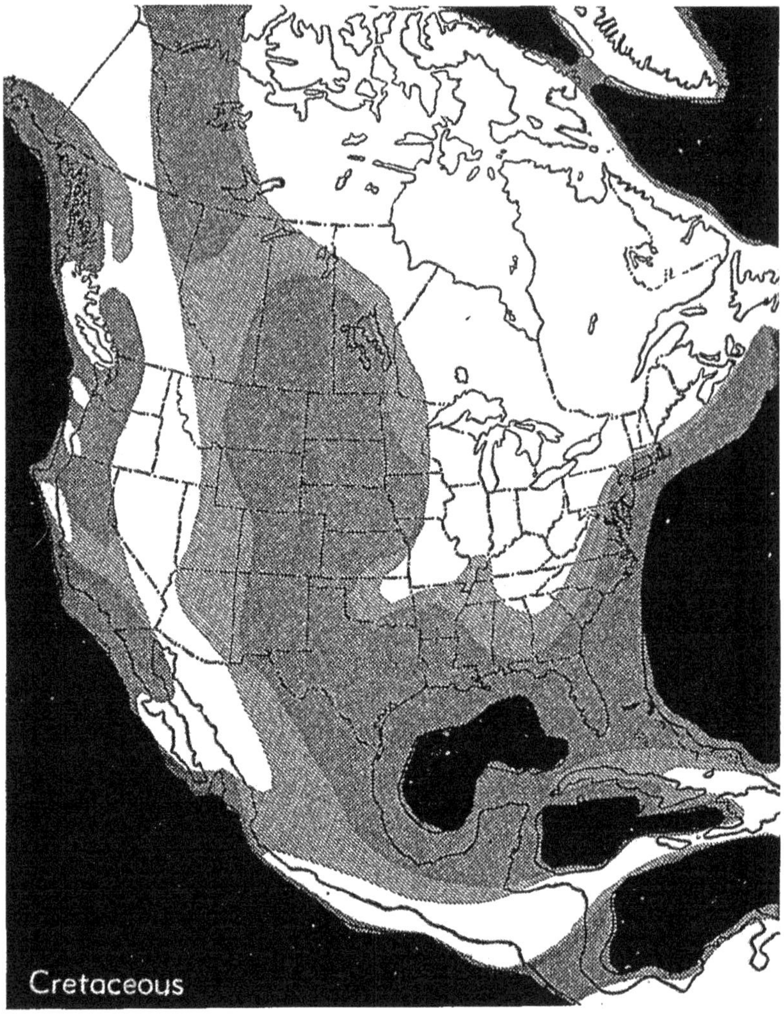

Figure 2-12. Typical of the Mesozoic Era is the Cretaceous map of North America. Continuously dry lands are shown in white, seas of progressively increasing depth are indicated by darkening shades of grey to black. Florida remained completely submerged throughout most of the Mesozoic Era. (Moore, Introduction to Historical Geology. McGraw-Hill Companies, 1949.)

developed between these embayments during the Cretaceous Period. The trough extended southwest across Georgia, through the Tallahassee area of Florida and into the Apalachicola depression. Geologists describe the trough as forming along a transition zone between marine and continental sediments. The trough was probably 20 to 30 miles wide and extended some 200 miles in length. Deposits of Cretaceous and Jurassic sediments eroding from Appalachia now fill the straits and embayments.

A most important geomorphic feature of the age was the progressive submergence of the southern extremities of the Floridan platform. This submergence possibly began in the Jurassic Period. The resulting submerged area has been designated as the south Florida embayment. Some investigators consider folding and downwarping the major cause of submergence. However, tensional faulting associated with the folding is recognized as a secondary process. The downward folded embayment plunges from a

Figure 2-13A. A probable Jurassic landscape of Florida before the state submerged beneath Mesozoic seas. While no fossil record of Florida at this time exists, it is assumed similar to areas lying to the north. This restoration shows several flying reptiles (pterosaurs) in the sky and trees; while near the ground early feathered birds (archacopteryx) with sharp teeth dive above the heads of tiny dinosaurs. (The Field Museum, #75017, Chicago.)

Figure 2-13B. A variety of life abounded in the Cretaceous seas which flooded Florida. Skeletal remains of these are found in various marine deposits. A large lizard (Mosasaur) which grew to a length of 50 feet is seen swimming near a large marine turtle. The hammerheaded pterandons probably roosted on land, but fed at sea. (The Field Museum, #66186, Chicago.

Figure 2-13C. Other Florida marine life included the long-necked, powerful, flippered plesiosaur shown swallowing a fish, and the porpoise-like ichthyosaur. (The Field Museum, #72379, Chicago.)

high near the Bahamas uplift into the depths of the Gulf Basin. The basin is approximately 800 miles long and some 300 miles wide. Sediments in the embayment attain thicknesses of over 15,000 feet. The submergence apparently advanced northward along the Florida Peninsula as indicated by marine beds of the Peninsular Arch. The remains of Florida's oldest known vertebrate were discovered in such Cretaceous deposits. The bones are believed to be those of a sea turtle. Late in the Cretaceous period the Peninsular Arch itself was inundated and covered with a relatively thin veneer of sediments. The advancement of the Mesozoic seas is shown in Figure 2-11.

During the Triassic and early Jurassic Periods, peninsular Florida was connected to a much enlarged North America. Appalachia had enlarged to include the middle portion of today's United States and Canada. Plants and vertebrates alike thus found easy access to the area.

Evergreens, pines, ferns, ginkgos, along with cycads (a palm-like plant) were the dominant types of vegetation. The remains of many such plants are found in coal-swamp deposits in the north. Modern oak, maple, hickory, and other types of flowering plants were only beginning to enter Cretaceous Florida at the time of its submergence.

The dominance of reptiles is the outstanding feature of Mesozoic life. They dominated the continents from the Triassic until the close of the Cretaceous, a period of 170 million years. The term dinosaur (terrible lizard) has been applied to these creatures. Dinosaurs appeared in a complex variety. Their size ranged from that of a mouse to massive. The largest, Brontosaurus, attained a length of 85 feet, stood 25 feet high, and weighed perhaps 50 tons. The Brontosaurus was an unarmored, semi-aquatic herbivore. Other large herbivores were well-adapted armored dinosaurs included the Ankylosaurus which resembled a huge armadillo and the Triceratops with three massive horns and shielded neck. Tyrannosaurus was by far the greatest of the carnivorous dinosaurs. This reptile stood 20 feet high, had an overall length of 50 feet, and weighed some 10 tons. Flying reptiles, birds, and small herbivorous and probably insectivorous mammals also roamed the land. No evidence of land-living dinosaurs has been found in Florida.

Different groups of dinosaurs were particularly adapted to life in the sea. Their body outlines were streamlined for fast movement and for propelling themselves through the water. Perhaps the most significant of these marine dinosaurs were (1) the ichthyosaurs or fishlizards whose body resembled that of a fish, but with an extremely elongated and pointed head for catching fish; (2) plesiocaurs that had a short broad body, a long neck, a small head, and an overall length of 50 feet; and (3) large turtles, growing up to 11 feet in diameter. (See figure 2-13A, B, C).

Obviously, the progressive submergence of Florida during the Cretaceous Period altered the major life forms from land dwellers to marine creatures. Numerous fossils of oysters and mollusks remain from the Mesozoic Period, but few remains of larger marine reptiles exist. This is probably due to unfavorable conditions for preservation, including periodic aeration and oxidation, short intervals of reemergence and erosion, and finally fairly complete submergence to open ocean conditions.

Science now believes it has determined the cause of the massive life extinction that ended the Mesozoic Era 65 million years ago. Satellite technology has located the site of a giant meteor or asteroid that belly-flopped along the Yucatan coast of Mexico. The energy released was equivalent to 75 million tons of TNT. The collision created a crater 150 miles in diameter and 24 miles deep. Rocks around the site were altered by heat and pressures.

An undersea shockwave traveling some 500 miles per hour formed in the Gulf, spreading out in all directions. As the shockwave was forced upward on the continental shelf, the energy was converted into an 800 foot high wall of water, called a **tsunami**. All marine life in its path was swept away, most to be killed by the pounding waves or buried in the sediment clouds created on the sea floor. The tsunami traveled inland hundreds of miles, sweeping everything in its path;

including dinosaurs, shellfish, and many other life forms. The shock waves also caused volcanos to break through the crust along fault lines. Ash from these eruptions joined the dust thrown into the air upon impact. The combined result caused years of darkness and "cosmic winter," in which the temperature of much of the earth dropped to near freezing conditions. The "deep freeze" was then followed by centuries of intense warming. Life forms (both marine and terrestrial) unaffected directly by the impact, disappeared in these atmospheric after effects, which apparently lasted for thousands of years. Scientists have located additional meteor impact craters of Cretaceous Age, which suggest that the era may have actually ended with a series of collisions similar to the one described. Some 75% of all species vanished at this period, and no dinosaur records exist after the end of the Cretaceous.

CENOZOIC PALEOGEOGRAPHY

During the early Tertiary Period (especially during the Paleocene and Eocene Epochs) Florida lay submerged in the deep ocean. The shoreline of this ocean extended across Georgia and Alabama, and circled northwestward up the Mississippi Embayment. Northwest Florida lay partially within the embayment described and received ample supplies of black mud contributed by tributary rivers. Marine limestone deposits accumulated over the remainder of the state. The boundary between deep-sea limestones and sandy and/or clay continental deposits shifted back and forth from time to time across northern Florida, producing an interfingering of the limestone and sediments. Several sediments appear to be deposited irregularly probably due to fluctuating sea levels.

The Ocala limestone which underlies most of Florida was laid down in an open, fairly shallow sea. The peninsula level was probably rising during the late Eocene Epoch of the Ocala deposition. By the following Oligocene Epoch, Florida probably stood above sea level.

Three anticlinal or uplift features are well documented from that time. These features include the Ocala Uplift, the Sanford High, and the Chattahoochee Anticline.

The three arches were eroded to near sea level conditions late in the Tertiary Period in a process called **peneplanation**. Gradual subsidence following peneplanation resulted in the formation of marshes, lagoons and swampy topography. These conditions were ideal for aquatic life, the formation of organic deposits, and the deposition of phosphates.

The deposition of phosphates during this period is particularly important to Florida, as it is now one of the state's primary mineral resources. The phosphates are believed to have originated from an upwelling, cold water current along the eastern continental shore. The phosphate-rich waters swept into the lagoons, where marine organisms selectively used the phosphates to construct their skeletal elements. Remains of these marine animals settled to the bottom, where they formed phosphate beds which intermingled with sands, clays, or limestones. These phosphate beds were later covered with additional sediments. Subsequent large scale solution of the limestone rocks concentrated the phosphate nodules into extensive "land-pebble" deposits.

As the advent of the Pliocene Epoch progressed, rising sea levels flooded the lower (southern) portion of late Miocene Florida and the Bahamas to the east. Two reef or island areas are believed to have remained intact. In the area of West Palm Beach-Miami a series of reefs probably became exposed to form a chain of cays and inlets similar to the Keys of today. To the west a second series of reefs existed near Immokalee; the two reef tracts partially enclosed a calm and shallow lagoon.

The Miocene climate was wetter and warmer than today's; this tropical climate spread as far north as middle Canada. Because the Florida Peninsula was submerged during the early Tertiary, fossil records reflect only deep sea life, particularly small-shelled snails called "**Foraminifera**," mollusks, enchinoidia or sand dollar-like creatures, and fish. However, the abbreviated peninsula during the Miocene time (middle Tertiary) was inhabited by a considerable variety of animals which replaced the dinosaurs as the predominant life form.

Recent paleographic studies of fossils indicate two million years ago the Florida coast was encircled by mangrove jungles. The mangroves and Caloosahatchee River, which drained the Peninsular, were home to crocodile-like creatures (up to 30 feet in length) and manatees. Grasslands and open savannas extended throughout the center of the peninsula and into the mainland of Georgia, westward through Texas and beyond. The grasslands were dotted with Tropical and sub-tropical vegetation, including oak, elm, breadfruit, buckhorn, camphor, and persimmon. Leaves of these trees feed giant ground sloth and mastodons (early elephants). Tall grasses provided food and cover for herds of camels, deer, three-toed horses, as well as rhinoceroses, dogs, weasels, giant armadillos, jaguars, and the newly discovered "Terror Bird" (Titanis) (See Figure 2-16, A-D).

This Terror Bird apparently migrated over the Middle American land-bridge from South America, where it found a suitable home in Central Florida. Titanis was surprisingly similar to Velociraptor and other predatory dinosaurs which lived millions of years earlier. It was probably six feet tall and weighed some 350 pounds. Skeletal studies suggest that the bird would stalk its prey lion-like through the tall grasses, attack at high speeds, and disable its prey with its giant beak. Titanis would then use its "hands" and 4-inch claws on its feet to kill its prey. Studies suggest this creature may have survived until 15,000 years ago. The remains of many of these Miocene (and later) creatures have been located in buried Ocala sinkholes and former swamp deposits where the animals were trapped and died.

The Pliocene Epoch was terminated by a period of extensive crustal warping throughout North America. The submerged area in the north-central part of Florida was warped upward to stand more than 300 feet above present mean sea level. The warping also contributed to our present broad embayment along the west coast between Clearwater and Apalachicola. Contemporaneous crustal warping lifted lands in Mississippi and Alabama while drowning the New England coastal plain, submerging the Hudson River Canyon. There was a drowning of other major eastern seaboard river valleys (i.e., Delaware and Susquehanna). Ultimately this deformation opened or widened Baffin Bay, Davis Straits, and the Greenland/Norwegian Sea. The expanded channel promoted an alteration of the Labrador current (cold) and North Atlantic Drift (warm water current), and indirectly altered major snowfall patterns. These changes are believed to have been the ultimate cause of the following Pleistocene Ice Epoch.

The cessation of the crustal deformation activities described mark a convenient timed event from which to date the beginning of the Pleistocene Epoch. The Pleistocene was marked by a series of sea level oscillations caused by the alternate withdrawal and return to the sea of great quantities of water. The waters were withdrawn in a frozen form and composed the continental ice caps that gave rise to the name "Great Ice Age." Studies indicate that during at least four Pleistocene ice ages, portions of North America, Europe, Asia, and South America were covered by perennial ice caps up to one or two miles thick. The extent of the North American ice sheets is shown in Figure 2-14. Each ice age persisted for thousands of years, although minor recessions of the ice sheets probably occurred during each glacial period. The four North American glacial stages have been named Nebraskan, Kansan, Illinoisan, and Wisconsin respectively. Each glacial stage was succeeded by a comparatively ice-free epoch lasting longer than its preceding glacial period. The warm interglacial stages have been labeled the Aftonian, Yarmouth, and Sangamon stages.

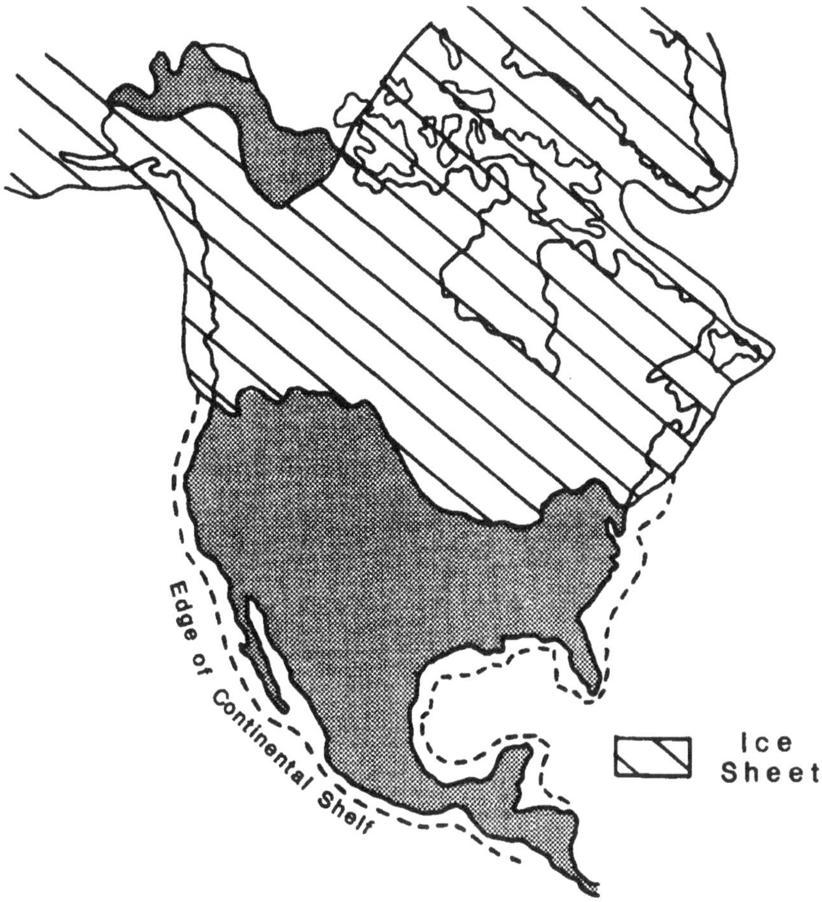

Figure 2-14. *Maximum extent of the North American ice sheet during the Pleistocene. The continental margin enlarged as shown due to lowering of sea levels.*

Although North American ice sheets extended only as far south as Kentucky and Kansas, the oscillations of sea level induced by their advance and retreat significantly influenced the topography of Florida. Today the Florida topography is marked by a series of terraces, which represent coastal surfaces of erosion or deposition created by Pleistocene and recent wave action. Each marine terrace represents the former bottom of a shallow sea floored with deposits of sand, clay, silt, and shells. Each terrace exhibits shoreline features such as relict beach ridges, lagoons, wave-cut scarps, and offshore and bay bars. The formation of most of the terraces is associated with the advance of Pleistocene seas during interglacial periods. Each successive submergence left larger and larger areas above the sea. Thus, all of dry land Florida was subject to erosion during each glacial stage, and increasingly larger areas remained above the wave action level during successive interglacial stages.

Two important marine processes which occurred during the Sangamon interglacial stage influence Florida's landscape today. The first of these was the growth of an extensive coral reef along the steeply sloping southeastern edge of the plateau. It has been found that coral will grow successfully in waters as deep as 120-150 feet. As time progressed, seas were even more supportive to the growth of what is termed the "Key Largo" limestone reef. The reef probably grew up to within 20 to 25 feet of the surface until its death, which was caused by the fall of sea level attending the Wisconsin glacial stage.

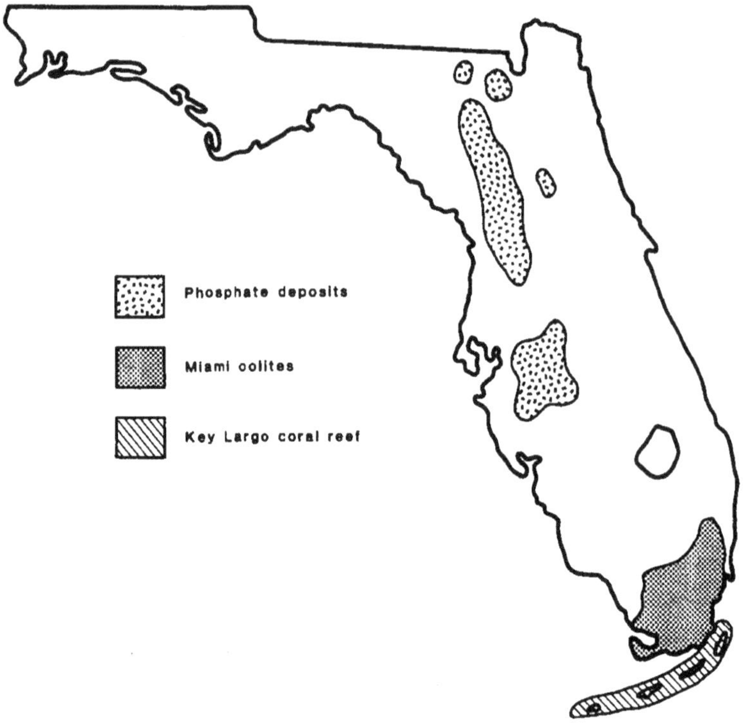

Figure 2-15. The map depicts Pleistocene accumulations of phosphates and Miami oolites. Key Largo corals and limestone appeared during the period along the southern tip of Florida.

In shallow waters between the Key Largo reef and coastal beaches, deposits of "Miami oolite" began to accumulate contemporaneously. The oolite formed as small spherules of calcium carbonate surrounding nuclei of sand grains or shell fragments. This limey oolitic ooze accumulated into thicknesses of up to about 30 feet. This oolite bar now forms the eastern rim of the Everglades and is the foundation upon which the cities of West Palm Beach and Miami are built. The locations of Miami oolites and Key Largo corals are shown in Figure 2-15.

Following early Wisconsin lowering of the sea level, portions of southern Florida were reflooded to a depth of 10 to 25 feet. This mid-Wisconsin recession constituted Florida's Pamlico seas, which surrounded a Florida shaped much like that of today. The peninsula was shorter, however, and terminated near Lake Okeechobee. A narrow strait, about five miles in width, separated the large Immokalee island from the peninsula. Today, this strait is occupied by the Caloosahatchee River. On the west coast an island on the Pinellas peninsula stood guard over the entrance to an enlarged Tampa Bay. A chain of large islands resembling those along today's Georgia and South Carolina coasts protected and elongated the lagoon which stretched from St. Marys to Sanford. Today the lagoon is occupied by the St. Johns River. Today, much of the sand along the St. Johns was probably blown from the low barrier islands, where it became protected from further erosion by roots and stems of water plants.

Importance of Terrace Studies

The study of terrace features helps us to better understand the processes shaping and affecting Florida. The study provided (1) an historical perspective of Florida's recent geology; (2) assistance in hydrologic problems; (3) a subsidence level indicator; and (4) assistance in locating mineral deposits. Terraces play an important role in hydrologic regimen of groundwater in many areas. The

Figure 2-16A. Wild hogs of the Miocene time were about 6 feet at the shoulder, but with a very small brain, suggesting a profound stupidity. On the right a Chalecothesenes nibbles from a tree. This mammal was horse-like with long claws, probably to dig roots. (The Field Museum, #71637, Chicago.)

Figure 2-16B. Example of a Plliocene rhinoceros, several of whose remains have been found in Florida sediments. This rhinoceros attained a height of about 4 feet, was short-legged, and primarily inhabited stream valleys, such as the Caloosahatchee of western Florida. (Neg. no. 35798. Painting by Charles R. Knight. Courtesy Dept. of Library Services, American Museum of Natural History.)

Figure 2-16C. Pleistocene flesheaters collect around the carcass of a grasslands herbivore (perhaps a horse, camel, or elephant). The carnivorous mammals of this period include the dog and cat families, such as the saber-tooth tiger seen snarling to chase away scavengers. (The Field Museum, #73799.1, Chicago.)

Figure 2-16D. Florida terror birds probably stalked their prey through tall prairie grasses, and when close, charged like a lion. Here the terror bird lunges for a frightened 4-horn antelope. (Steven Kirk / © 1997. Reprinted with permission of Discover Magazine.)

relationship of terrace altitude to adjacent lands and to size sorting of underlying sediments directly affects the occurrence and movement of groundwater. Such an effect is significant on the piezometric surface, which represents the level to which artesian or pressurized groundwater will rise in tightly cased wells. The configuration of the shallow artesian aquifers generally reflects land surface contours, particularly in such areas as Duval, Escambia, and Santa Rosa Counties. In those areas where well-sorted permeable sands underlie the terrace, groundwater recharge occurs by rainwater infiltration into the shallow sand aquifer.

Mineral deposits associated with terraces, shore, and near-shore features are of economic importance. Vernon (1943) states that phosphate deposits are confined beneath three terraces. The three terrace flats lie near 100, 150, and 212 feet above sea level, respectively. MacNeil (1949) found outer coastal bars of the Pleistocene are important economic sources of heavy minerals. These include brick and cement clays, china clays, kyanite, titanium deposits, and rare earth monazites.

In Florida, the higher the shoreline the greater its age, and hence the greater its modification probability by solution of underlying limestone. Relict beach ridges, which remain level along their length, indicate little or no movement. Conversely, variations in terrace elevation along its length is evidence for tectonic action or solution-induced subsidence. With the absence of lakes, recognized depressions, or sinkholes, discontinuities of crest heights along a terrace may be the only significant surface indicator of subsurface subsidence.

Recently the Cutler archaeological site in an old Dade County sinkhole has yielded direct evidence of Ice Age Florida. The site indicates that the South Florida climate was dry and cold with occasional snowfalls, similar to present day winters of North Carolina. Cave bones indicate woolly mammoths, bison, cougars, and wolves the size of shetland ponies once ran freely through the state, while condors circled in the Florida skies. At this period man was hunter, living in caves for warmth, cooking his kill over fires, and learning from the land.

Recent Geomorphic Activities

The term "Recent Period" applies to the history of North America since the end of the Wisconsin glacial stage of the Pleistocene. At that time, the melting of massive glacial ice accumulations flooded the oceans with melt water and caused them to rise approximately 300 feet. The level has remained essentially constant since that time. This final rise of sea level flooded many stream valleys which had been carved into channels below the general level of the land.

Marine conditions along the Florida coast are quite similar to those during the high sea level stages of the Pleistocene. Atlantic long shore currents are sweeping sands south as far as Miami Beach, while Gulf currents similarly wash the west coast. Florida Bay and adjacent marine areas are slowly filling with an ooze somewhat like that of Miami oolite, and a new coral reef is developing off the Key Largo limestone.

The various geographic processes currently molding and sculpturing the Florida lands are described in succeeding chapters.

REFERENCES

Arrington, D. Albrey, (1995), Restoration Ecology Journal of the Society of Ecological Restoration Vol. 3, No. 3, 9/95

Brown, M.W. "New Clues to Earth's Largest Extinction," New York Times Orlando Sentinel, January 24, 1993, P. G-7.

Carpenter, R. H. (1970), "Metamorphic History of the Blue Ridge Province of Tennessee & North Carolina," *Geological Society of America* Bull. Vol. 81, pp. 749-762.

Cooke, C.W. (1945), *Geology of Florida:* Florida Geol. Survey Bull. 29.

Dietz, R. S. & J. C. Holden (1970), "Reconstruction of Pangaea: Breakup and Dispersion of Continents, Permaian to Present," *Journal of Geophysics Research,* Vol. 75, pp. 4939-4956.

D'Oliverira, S., "Here It Comes," *Orlando Sentinel* October 20, 1996, p. 7, Discovery Channel, "3 Minutes to Impact," program February 19, 1997.

Emmons, W. H., 1. S. Allison, C. R. Stauffer, C. A. Thiel (1960), Geology: Principles and Processes, McGraw-Hill Book Co., New York, p. 491

Flazel, C. T. (1986), "Asteroid Impacts, Sea Floor Sediments, and Extinction of the Dinosaurs," *Sea Frontiers* Vol. 32, pp. 169-178. Forest, H. E. (1935) *The Atlantean Continent Its Bearing Upon the Great Ice Age and the Distribution of Species* 2nd ed., London.

Hatcher, R. D. (1972), "Developmental Model of the Southern Appalachians," *Geological Society of America Bull.,* Vol. 83, pp. 2735-2760.

Healy, H. G. (1975), Terraces and Shorelines of Florida: Florida Geol. Survey M.S. 71.

Knight, C. R., Chicago Natural History Museum.

MacNeil, F. S. (1949), Pleistocene Shorelines in Florida and Georgia: USGS PROF. PAPER 221-F, pp. 91-107.

Marsh, 0. T. (1966), *Geology of Escambia and Santa Rosa Counties, Western Florida Panhandle:* Florida Geol. Survey Bull. 46. Moore, R. C. (1949), *Introduction to Historical Geology* McGraw-Hill Book Co., New York, p. 582.

Murray, G. E. (1961), *Geology of the Atlantic and Gulf Coastal Province of North* America: New York, Harper & Brothers, p. 692.

Puri, H. S. & R. 0. Vernon (1964), *Summary of the Geology of Florida and a Guidebook to the Classic Exposures,* Special Publication 5, Florida Bureau of Geology.

Sawkins, F. J., G. C. Clement, D. G. Darby, & G. Rapp (1974), *The Evolving Earth* Macmillan Publishing Co., New York, p. 477.

Sullivan, W. (1984), "Geologists add more Pieces to a Global Jigsaw Puzzle," *Discover* pp. 66-75.

Watkins, J. S. (1971), "Sea-floor Spreading 196 Crustal Convergence Model for the Southern Appalachians," Geological Society of America, abs. with Programs (Southeastern Section), Vol. 3, No. 5, pp. 357-358.

Watkins, J. S. (1970), "Evidence of Middle Paleozoic Sea-floor Spreading in the Southern Appalachians," Am. Geophys. Union Trans., Vol. 51.

Zimmer, C. (1997), "Terror Take Two," *Discover,* pp. 68-74, June 1997

SUMMARY OF CHAPTER 2

The face of Florida is continually changing. The paleogeography of Florida closely relates to the concept of plate tectonics and the collision of the North American, African, and South American plates. In the early stages of the plate collision, Florida began as a series of Paleozoic volcanic islands which formed near the subduction zone. Life forms on these islands were probably similar to that of Appalachia to the north, although no fossil remains of the Paleozoic era still exist. The islands eroded and reformed several times during this age, but late in the era Paleozoic orogenies uplifted the lands again to form the Peninsular Arch from the peninsular sediments.

During the Mesozoic Age the plates decoupled and Florida was severely fractured in the process. Large embayments formed along the northern portion of the plateau, separating Florida from the North American mainland. However, later sediments re-established the land connection. Life forms consisted of dinosaurs which fed upon one another and lush tropical vegetation. Toward the middle of the era southern Florida subsidence increased significantly, and before the end of the Mesozoic the entire peninsula lay submerged.

The submergence continued well into the Cenozoic era, during which time significant accumulations of marine limestones were deposited along the platform. Compressive forces from the west eventually resulted in the formation of at least three anticlines or uplifts. These uplifts elevated the limestones and exposed them to a process of erosion. During the Pleistocene (a late period of the Cenozoic) four glacial/interglacial periods alternately lowered the sea level, (with exposure of the entire Florida Continental Shelf) and re-inundated the platform. The cycles of inundation created a series of marine terraces on the Florida peninsula-terraces, which still exist today. The terraces continue to exert important geological influences on the state's natural resources.

Cenozoic life forms of the peninsula consisted primarily of mammals; signs of man's activities are also suggested from this era.

Name ______________________ Section ____________ Date____________________

REVIEW OF CHAPTER 2

Part A

1. The oldest rock type underlying Florida contains no advanced types of invertebrate life. These rocks formed during the ____________ era.
2. Forces exerted by ____________ believed to move tectonic plates about the globe.
3. When tectonic plates collide trenches are produced by a process while mountains result from a crumpling process termed ____________.
4. During the Paleozoic Era the eastern seaboard of the North American plate was characterized by
   a) Pangaea
   b) Continent of Appalachia
   c) Avalon
   d) Mid-Continental Sea
5. During the Paleozoic Era the climate of Florida was probably
   a) Arid with little seasonal variations
   b) Alternating wet-dry due to changes in winds patterns
   c) Cool due to polar sea currents
   d) Humid due to warm surrounding seas
6. Florida's fossil record suggests the life forms of the Paleozoic included
   a) amphibians
   b) reptiles
   c) early mammals
   d) all of the above
7. Florida's Peninsular Arch probably formed
   a) in conjunction with the Appalachian orogonies/uplift
   b) during decoupling or separation of the African and North American plates
   c) due to tensional stresses in Mesozoic age rocks
   d) due to sea floor spreading
8. Mesozoic folding and faulting in south Georgia and northern Florida produced
   a) Suwanee Straits
   b) South Florida Embayment
   c) Okefenokee and Appalachicola Embayments
   d) all the above
9. Most of Florida is underlain by Cenozoic (Eocene) deposits of
   a) interfingered black muds and marine sands
   b) eroded sediments from the Ocala Uplift
   c) glacial deposits
   d) limestones laid down in shallow seas

10. Pleistocene ice sheets influenced Florida's topography by
    a) scrapping away surfacial layers
    b) increasing stream erosion
    c) increasing coastal erosion due to rising sea levels
    d) promoting the formation of coastal terraces

Part B

1. How is the formation of the Florida Plateau related to plate tectonics? What was Florida's Paleozoic form?

2. Why does no surface expression of the north Florida trench remain today?

3. How was Florida's Peninsular Arch formed? How is this arch related to features found in Georgia and the Bahama Islands?

4. What has been the effect of plate decoupling on Mesozoic Florida?

5. Why did Florida begin to submerge during the Mesozoic? How extensive was the submerging?

6. How were Florida's phosphate deposits formed?

7. What crustal changes occurred in Florida just before the Pleistocene glaciation?

8. List four reasons why the study of terraces is important.

9. How are terraces related to acquifer recharge areas?

10. How is the age of relict beach ridges related to subsidence in the central Florida area?

11. What terranes are recognized in the southern portion of the North American plate?

12. What is the recent theory to explain the extinction of the dinosaurs?

CHAPTER 3

Climate—Florida's Most Important Resource

INTRODUCTION

Climate is frequently referred to as Florida's most important resource because it influences so many of the other state resources and so many facets of the economy. Agriculture is an important resource but because of the large areas of almost sterile, sandy soil, growers can be successful largely because they can grow crops during the mild winter season when it is too cold for crops in most of the United States. Climate also plays a large role in the tourist industry because the southerly location of the state affords an escape from the cold, frigid winters in the more northerly latitudes of the United States.

Climatologists and meteorologists differentiate between weather and climate. The term weather implies the hour-to-hour and day-to-day conditions of the environment—involving precipitation, temperature, humidity, pressure, and winds. Each of these items is variable and subject to change. Geographers are more interested in the long-term averages of weather conditions, especially temperature and precipitation, so the main emphasis in this chapter will be on climate.

Some speak of climate as "average-weather," but a more satisfactory way of thinking of "average weather" is the frequency of the occurrence of the precipitation and temperature values as well as the average of these values. The average value is known as the mean. A very important type of frequency is the number of occurrences of a certain deviation or departure from the mean. For example, the mean annual temperatures of Miami, Florida, and Yuma, Arizona, are both approximately 75°F (24°C); yet, the monthly means from which the annual means were computed are very different. Miami's greatest deviation from the mean annual temperature is 9°F (5°C) in January and 7°F (4°C) in July and August, compared to 20°F (11°C) deviation for both January and July in Yuma, Arizona.

CONTROLS OF CLIMATE

Weather varies from day to day and climate varies from place to place because of variations in the amount, intensity, and areal distribution of the several climatic elements, especially temperature and precipitation. Variations in weather and climate throughout the world exist because of certain natural phenomena known as climatic controls. They are (1) location, size, and shape; (2) topography; (3) planetary wind and pressure systems; and (4) ocean currents.

Every place has a specific location designated by latitude and longitude. Of the two coordinates, latitude is the more important in regard to climate. The term latitude specifies how far from the equator the area is located. This in turn, determines the angle at which the sun's rays reach the earth and the length of the day. Because the axis of the earth is tilted at an angle of 23½°, no place on the earth further north or south of 23½° latitude can have the sun's rays strike the ground perpendicularly. In other words the sun is never directly overhead, north or south, of the 23½° parallel. The quantity of heat per unit area is greater where the sun is directly overhead than where the sun strikes the earth at an angle of less than 90°.

In theory, the longitude of a particular area would not be important if the earth's surface was identical over the entire globe. For example, if water of the same depth and density covered the earth's surface, the latitude of an area would be the dominant control of climate, and all areas would have the same climate. However, because of variations in other controls this is not true.

Size is important in controlling climate because the larger the area, the further the interior locations are from the ocean and the more continental the climate will be. This results from the difference in rates of absorption of solar energy between land and water. Land absorbs and loses heat five times faster than does an equal volume of water. Large land masses such as central Asia are so far from the sea that the climate is not tempered by the cooler water in summer and the warmer ocean in winter. Such climates are called continental climates. Small masses of land such as Iceland, completely surrounded by a large expanse of ocean, have climates much more moderate than their latitude would indicate. These are termed marine climates.

Shape also influences the climate of a particular location. A square-shaped region, surrounded by water, would have interior locations farther from the sea than would a narrow rectangular region of the same area. This in turn would result in the latter region being influenced more by marine conditions. Florida's long, narrow peninsular shape makes it possible for all places within the state to lie within 70 miles of the warm ocean or gulf, giving it a relatively cool summer and a moderate winter.

The topography of an area may affect the climate in two ways. If the region is dominated by highlands, the climate will be cooler than it would be if it were located near sea level at the same latitude. This is because there is normally a decrease in temperature of about 3.5°F (1.9°C) for each 1,000 feet increase in elevation. This is known as lapse rate. The second effect of topography involves the distribution of mountains and lowlands, Mountains located parallel to and near the coast act as a barrier to moisture-laden winds blowing in from the ocean. This is particularly important if the mountains are high and lie across planetary wind belts, as do the Sierras in California. They cut off so much moisture that Nevada, to the east of the mountains, is practically a desert. On the windward California side, the air that is forced up over the mountains releases much rain in summer and snow in winter. This type of precipitation is known as orographic precipitation. See Figure 3-1. If a long, linear range of high mountains extended along the east coast of Florida from Miami to Jacksonville, the interior of peninsular Florida would be much drier than it actually is. Conversely east of the mountain range, along the coast, the rainfall would be much higher than at present.

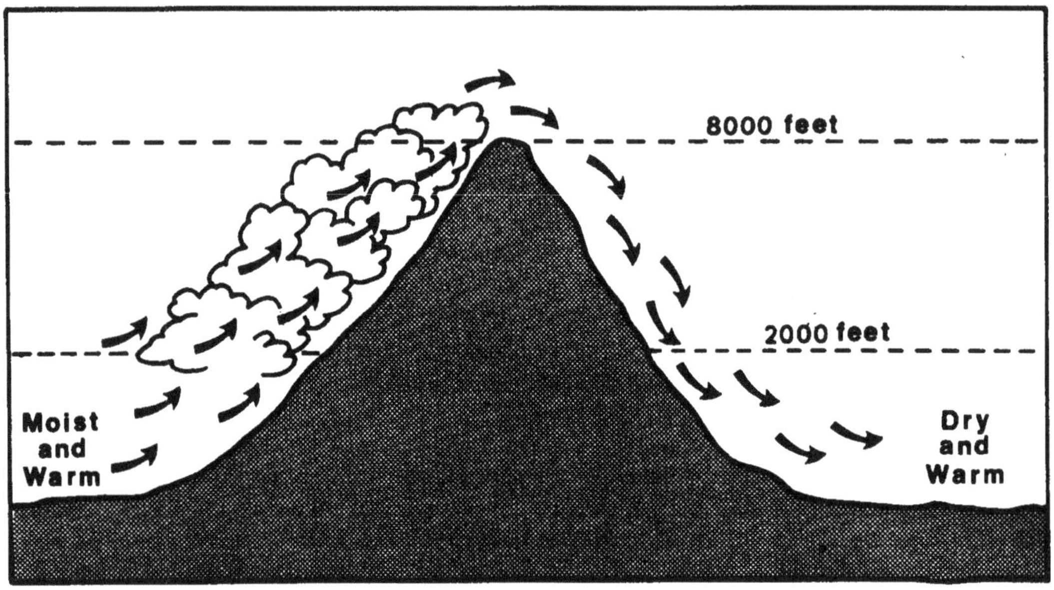

Figure 3-1. Orographic precipitation. As warm, moist air is blown upward over the windward side of a mountain, it may contain sufficient moisture to become saturated when cooled to its dew point. This can result in rain on the windward side of the mountain.

Winds are horizontal movements of air resulting from differences in pressure between two air masses or rising air within the same air mass. Any condition that reduces pressure, such as convectional lifting, will cause air to flow toward the region of rising air, the velocity of the wind depending upon the pressure gradient. Changes in wind direction and velocity bring about changes in temperature and precipitation, which together largely determine the character of wind and climate.

It is also important whether the wind originates from the poles or from the equator because of the temperature induced. Winds are named for the direction from which they come and bring temperature and moisture conditions characteristic of the region of origin. In the northern hemisphere, winds from the north are usually colder than winds from the south.

Winds originating over water also bring different temperatures and moisture conditions than do those from landmasses. On a small scale these are known as land breezes and sea breezes. On the continental scale they are known as monsoons, a term from Arabic, meaning reversal. During the winter the winds blow outward from a high pressure area toward a warmer ocean. During the summer the reverse is true. The land is warm, causing the air to rise. Cooler, heavier air from the adjacent ocean flows toward the lower pressure region vacated by the rising air.

If mountains are near the coast, heavy precipitation results from orographic lifting. India is a classic example of this monsoon pattern of winds. In Florida, where the terrain is relatively flat and the land mass relatively small, the precipitation in summer is less that coastal India.

From the previous examples it can be seen that wind transports heat from lower to higher latitudes and is the principal agent in maintaining other latitudinal heat balance of the earth in spite of unbalanced conditions of radiation between low and high latitudes. Winds also transport moisture from the oceans, rivers, and lakes to regions where local ground moisture is scarce. The two major types of pressure distribution are high and low pressure. The rate and direction of change in pressure is called pressure gradient or barometric slope. When an air mass from a high pressure region meets an air mass from a low pressure region they form a boundary called a front. The air masses do not mix. Instead, air from the high descends, and the warmer, moist air from the low rises and cools, forming clouds, unsettled weather conditions, and frequently precipitation. This type of precipitation is called frontal or cyclonic precipitation.

GENERAL ATMOSPHERIC CIRCULATION

Earlier in this chapter, under "Controls of Climate," the effects of latitude were discussed. Generalized wind circulation and atmospheric pressure patterns as a function of latitude are shown in Figure 3-2. In the lower latitudes, during the daylight hours through the year, the position of the sun is high in the sky and the number of hours of daylight changes much less than is true of more poleward locations. As a result of the solar insulation, warm air rises, producing areas of low pressure and converging air. Such low pressure cells are called depressions, cyclones, or merely lows. A low pressure occurs throughout the year in the vicinity of the equator in a belt called the Intertropical Convergence Zone (ITCZ), more commonly known as the "doldrums" or equatorial low.

The ITCZ produces a difference in the density of the air as compared with regions farther from the equator and results in a global circulation of air in the earth's troposphere.* Much of the energy required to maintain the global circulation comes from the tropical oceans where evaporation is high and large amounts of latent heat are available to warm the air.

Straddling the ITCZ are regions of strong winds which converge upon the zone in the form of the trade winds. The trade winds

* The Troposphere is that portion of the earth's atmosphere extending from the surface of the earth upward to an elevation of approximately 10 kilometers or 33,000 feet.

originate in high pressure cells, at latitudes of approximately 30°N or S, known as the subtropical highs or horse latitudes. The difference in pressure between the subtropical high pressure cells and the ITCZ is known as pressure gradient. The pressure gradient between the horse latitudes and the ITCZ produces a continuous year-round wind belt known as the trade winds. In the Northern Hemisphere the belt is referred to as the Northeast trade winds. And in the Southern Hemisphere as the Southeast trade winds. Because the earth spins on its axis from west to east, winds are deflected to the right in the Northern Hemisphere and the left in the Southern Hemisphere. This deflection of winds is known as the Coriolis effect and results in the trade winds approaching the ITCZ from the northeast and southeast rather from due north and due south respectively.

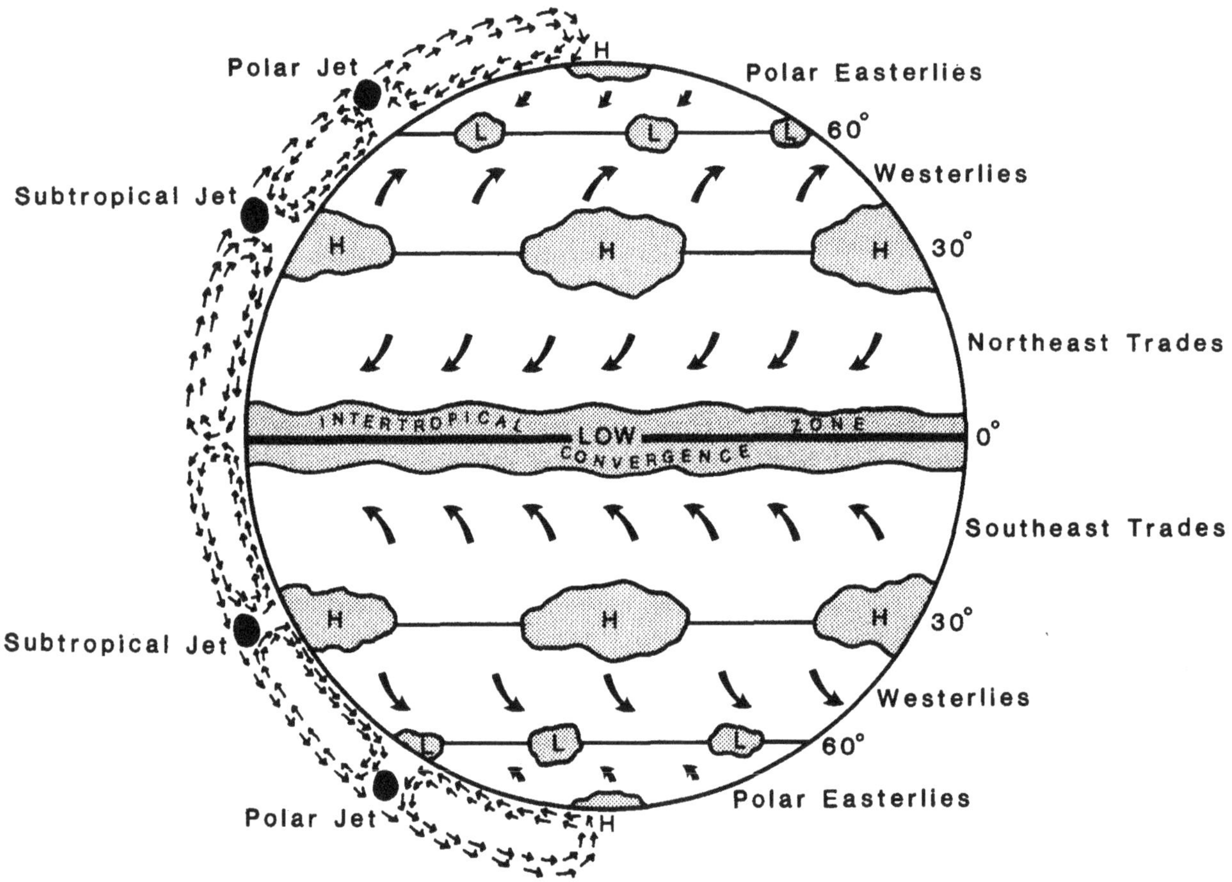

Figure 3-2. Generalized pressure pattern and wind circulation. The Intertropical Convergence Zone is the birth place of the global air circulation, and the Coriolis effect alters the direction of the wind flow.

The descending air forming the subtropical highs has its origin in the warm, moist air that rises over the ITCZ. When the air has risen to a height where its density equals the density of the surrounding air, it begins to move horizontally poleward until it reaches a latitude of approximately 30°N or S. At this latitude the air has cooled and accumulates before subsiding and diverging near the surface of the earth. Because the surface of the earth consists of continents and oceans, and because land and water heat and cool at different rates, the subtropical highs consist of a series of cells rather than a continuous belt around the earth. In Florida, the high pressure cell having the most influence upon its weather is the Azores-Bermuda high, which hangs off the coast of northeast Florida and is responsible for heat waves in late spring and summer.

Only part of the descending air that constitutes the subtropical highs flows toward the equator as the trade winds. The remainder of the diverging air flows from the high pressure cells toward the poles to form the prevailing westerlies. The Coriolis effect again deflects the wind to the right in the Northern Hemisphere and to the left in the Southern Hemisphere. Although generalized diagrams of the global circulation show the prevailing winds from the southwest and northwest in the Northern and Southern Hemispheres respectively, winds may blow from any direction. However, the winds blow from some variation of west most of the year. The prevailing westerlies are zones of cyclonic storms which move from west to east over the North and South American continents, especially during their respective winter seasons. On the North American continent the storms find their way into Florida, resulting in the cold spells of two or three days duration that characterize the winters of Florida.

All of the air that ascended in the ITCZ does not subside in the vicinity of the 30°N or S. Some of the air continues to flow to the poles where it descends and diverges to form the polar highs. The pressure gradient causes the air to move toward the equator, while the Coriolis effect results in winds blowing from the Northeast in the Northern Hemisphere and from the Southeast in the Southern Hemisphere, These strong, cold winds are known as the polar easterlies. In the vicinity of 60°N or S the cold, dry air of the polar easterlies meets the warmer air of the prevailing westerlies, resulting in the polar front, frequently called the subpolar low.

CLIMATIC CONTROLS AFFECTING FLORIDA

The previous discussion of climatic controls was included in the text for those students who have not had an introductory college geography course. Keeping the four controls in mind: (1) location, size, and shape; (2) topography; (3) wind and pressure systems; and (4) ocean currents, how do they influence the climate of Florida?

Florida is located at the southeastern tip of the North American continent, extending from about 24° North latitude to 31° North latitude. This places Florida during part of the year in or near the belt of the subtropical high, which in turn is the source of the northeast trade winds and the prevailing westerlies. Because the wind belts shift north or south each year, lagging after the sun, the season will determine which wind belt zone is nearest Florida and will influence the climate in conjunction with the location of the upper level jet stream and the climatic conditions to the north and west of the peninsula. The warm waters of the Atlantic Ocean and the Gulf of Mexico which surround the narrow peninsula on three sides moderate the cold continental air of winter and supply moisture for the summer convectional showers.

Florida's climatic pattern shows a monsoonal effect resulting from its location at the southeastern corner of a large land mass. Continental influences predominate during the cooler winter months while maritime influences control the summer regime. As a result, temperature contrasts throughout the state are greater during the winter than during the summer and wind direction changes from generally northwest in winter to southeast in summer. The reversal of the wind direction seasonally on a large scale is duplicated daily on a local scale as winds blow from the seas to land during the day and from land to sea during the night. These local changes in wind direction result in sea and land breezes respectively. (Figure 3-3)

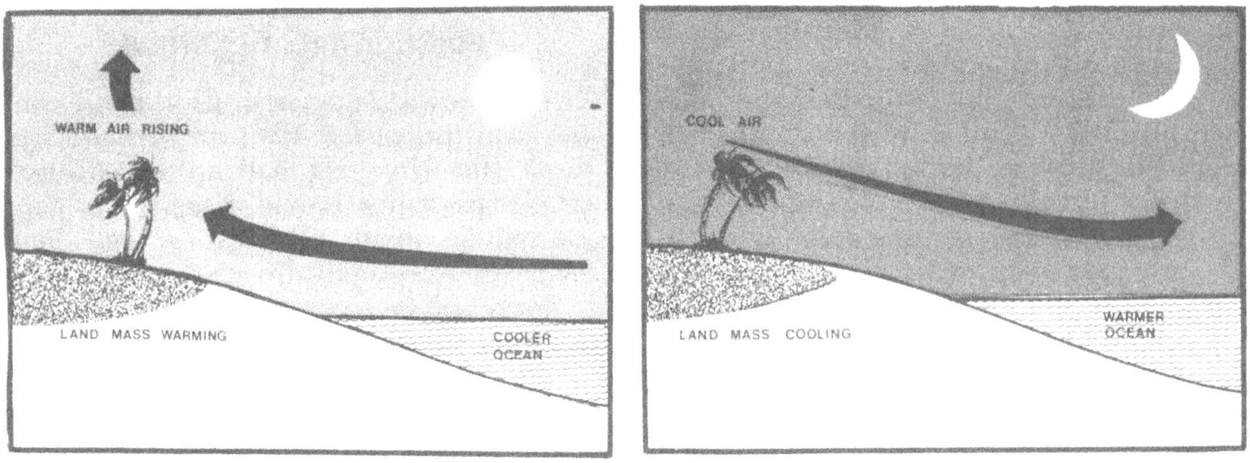

Figure 3-3. Coastal locations have changes in wind direction during the day and night. Left: During the day sea breezes blow from the cooler ocean toward the warmer land. Right: At night the wind direction is reversed when the land cools more quickly than the ocean and wind blows from the cooler land to the warmer ocean

TEMPERATURE

Mean annual temperatures for Florida range from the upper 60s°F in the northern section to the middle 70s°F along the southern coast of the peninsula to nearly 70°F (25.5°C) at Key West. Mean averages during the summer months are relatively uniform throughout the state—81°F to 82°F (27°C to 28°C). During the coolest months temperatures average about 13°F (7°C) lower in the north than in the south. July and August have the highest means in all areas of the state, while December and January averages are the lowest in the northern and central sections. January and February usually are the coolest months in the Keys. The lower January/February means illustrate the effects of the latent heat of the surrounding water. It takes approximately one extra month for the adjacent water of the ocean to cool, resulting in lower temperatures later in the winter season (See Table 3-1).

Extreme heat waves, characteristic of continental regions, are felt occasionally but in a modified form over the northern interior. The highest temperature ever recorded in Florida was at Monticello, in Jefferson County near the Georgia border on June 29, 1931 when an official United States Weather Bureau thermometer recorded 109°F (42.7°C). Usually heat waves arise when the Azores-Bermuda high lies too close to the east coast of Florida, preventing convectional thundershowers from developing. The lack of cloud cover and the loss of cooling from rain permit the heat to build up and high temperatures to result. Thundershowers, which on the average occur on about half of the summer days, are frequently accompanied by a rapid 10°F to 20°F (5.5°C to 11.1°C) drop in temperature, resulting in comfortable weather for the remainder of the day. As most of the large-scale wind patterns affecting Florida in the summer have passed over water surfaces, hot drying winds seldom occur. Dry winds do occur occasionally over south Florida when a high pressure cell develops in the Gulf in the vicinity of Ten Thousand Islands. The air heats as it descends, diverges, and blows across the Everglades without absorbing much moisture. By the time the air reaches Miami and Fort Lauderdale it is a hot, dry wind laden with mosquitoes. Despite these occasional hot spells, the highest official temperature ever recorded in Miami was 100°F (38°C) as compared with 109°F in Monticello. Not only are the maximum temperatures lower in south Florida, they are also of shorter duration.

The lowest temperature ever recorded in Florida was an unofficial -2°F (-18.8°C) at Tallahassee on February 13, during the freeze of 1899. Old-timers still speak of the "Big Freeze" of 1894-95, which killed most of the citrus trees in northeast Florida. Cold spells are listed in Table 3-2. A careful study of these severe cold spells indicate that they do not appear to follow any definite pattern. The time between cold spells ranges from as much as 22 years to as little as one year. The most severe freeze since the freeze of 1899 occurred in December, 1962. During this freeze the temperature dropped to 5°F (-15°C) on December 13 at De Funiak Springs and Crestview in northwest Florida. This was the lowest official temperature recorded during the 20th century, but not much lower than the 8°F (-13°C) registered at Tallahassee or the 9°F (12.7°C) recorded at Center Hill in Sumter County on January 13, 1981. It is interesting to note that Marianna, 55 miles east of De Funiak Springs, holds the record for the most days with temperatures below freezing—46 days during the winter of 1917-18. Key West, at the southern extremity, has never recorded a temperature below 41°F (5°C), which was reached in January 1886.

Table 3-1. Climatic Data for Selected Stations

Fort Myers			Jacksonville			Key West			Miami		
M	T	P	M	T	P	M	T	P	M	T	P
J	63.5	1.64	J	54.6	2.78	J	70.7	1.67	J	67.2	2.15
F	64.7	2.03	F	56.3	3.38	F	71.6	1.85	F	67.8	1.95
M	68.5	3.06	M	61.2	3.56	M	74.6	1.36	M	71.3	2.07
A	73.3	2.03	A	68.0	3.07	A	78.1	2.17	A	73.0	3.60
M	77.7	3.99	M	74.3	3.22	M	80.9	2.31	M	78.0	6.12
J	81.1	8.89	J	79.2	6.27	J	83.5	4.55	J	81.0	9.00
J	82.5	8.90	J	81.0	7.35	J	84.6	4.11	J	82.3	6.91
A	82.8	7.72	A	81.0	7.89	A	84.7	4.47	A	82.9	6.72
S	81.6	8.71	S	78.2	7.83	S	83.2	7.34	S	81.7	8.74
O	76.4	4.37	O	70.5	4.34	O	79.6	5.57	O	77.8	8.18
N	69.4	1.31	N	61.2	1.79	N	75.1	2.67	N	72.2	2.72
D	64.8	1.30	D	53.4	2.59	D	71.3	1.52	D	68.3	1.64
Yr.	73.9	53.95	Yr.	68.4	54.47	Yr.	78.2	39.95	Yr.	75.5	59.80

Orlando			Pensacola			Tallahassee			Tampa			West Palm Beach		
M	T	P	M	T	P	M	T	P	M	T	P	M	T	P
J	60.3	2.28	J	52.1	4.37	J	52.6	3.74	J	60.4	2.83	J	65.5	2.60
F	61.5	2.95	F	54.8	4.69	F	54.8	4.77	F	61.8	2.86	F	66.1	2.60
M	65.9	2.48	M	59.9	6.31	M	60.3	5.93	M	66.0	3.89	M	69.8	3.32
A	71.3	2.72	A	68.1	4.99	A	67.9	4.07	A	72.0	2.10	A	73.9	3.75
M	76.4	2.94	M	75.2	4.25	M	74.8	4.04	M	77.2	2.41	M	77.5	3.17
J	80.2	7.11	J	80.6	6.30	J	80.0	6.62	J	81.0	6.49	J	80.5	8.14
J	81.4	8.29	J	81.8	7.33	J	81.1	6.92	J	81.9	8.43	J	81.9	6.52
A	81.8	6.73	A	81.8	6.67	A	81.1	6.89	A	82.2	8.00	A	82.3	6.91
S	80.1	7.20	S	78.3	8.15	S	78.1	6.64	S	80.8	6.35	S	81.5	9.85
O	74.3	4.07	O	70.0	3.13	O	69.3	2.93	O	74.7	2.34	O	77.2	8.75
N	66.6	4.56	N	59.5	3.37	N	58.9	2.81	N	66.8	1.79	N	71.0	2.48
D	61.3	1.90	D	53.8	4.66	D	53.2	4.22	D	61.6	2.19	D	66.8	2.21
Yr.	71.8	51.21	Yr.	68.0	64.22	Yr.	67.7	61.58	Yr.	77.2	49.38	Yr.	74.5	62.06

Source: NOOAA, Environmental Data Service, National Climatic Center, Asheville, N.C. *Climate of Florida.*

T = Temperature in ° F   P = Precipitation in inches of rainfall

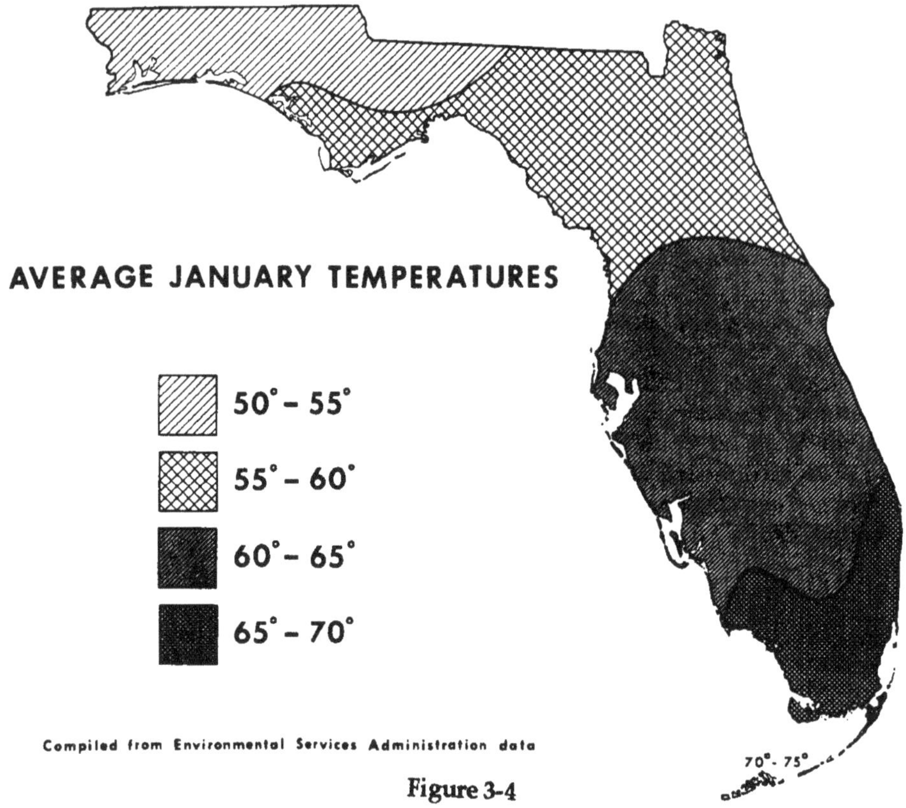

Figure 3-4

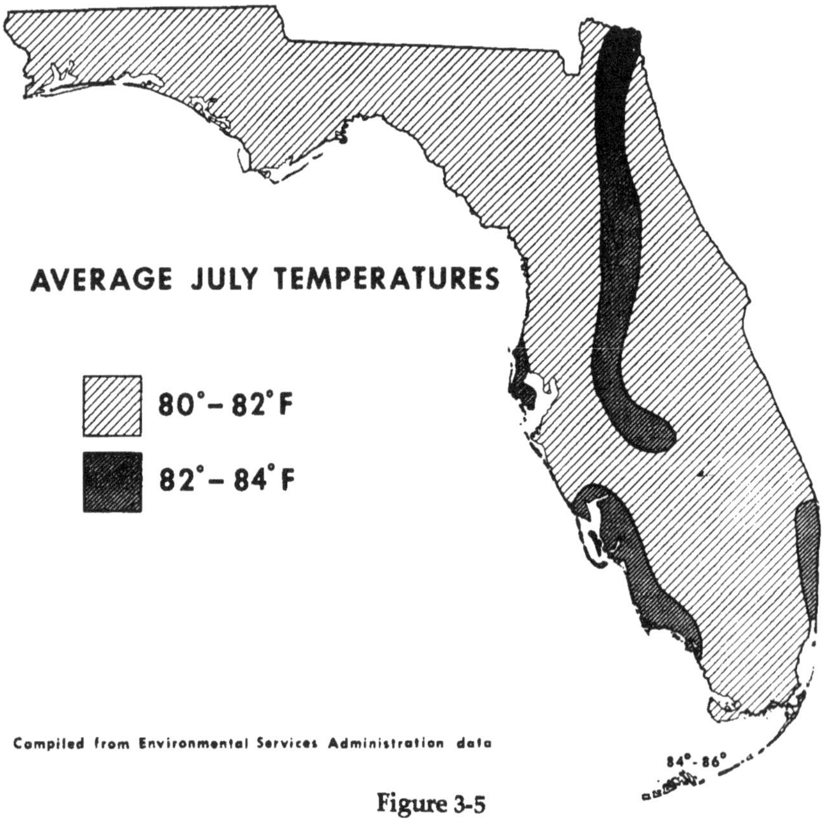

Figure 3-5

Table 3-2. Selected Severe Cold Spells in Florida

January 1766	December 1909	January 1971
February 1835	February 1917	February 1973
February 1857	January 1928	January 1977
December 1870	December 1934	January 1981
December 1880	January 1940	January 1982
January 1886	February 1947	December 1983
February 1895	December 1957	December 1984
January 1898	February 1958	January 1985
February 1899	December 1962	January 1989
January 1905	January 1966	February 1989
December 1906	November 1970	

Source: NOAA, Environmental Data Service, National Climatic Center, Asheville, N.C., *Climate of Florida*.

Two mechanisms produce the freezing conditions that often occur during Florida winters. The first, radiation frosts, result when large radiant heat losses cool the ground, plants, and surface air to freezing or near freezing levels. This may result in frost covered vegetation under proper humidity conditions. The colder air drifts downhill, following the curves and contours of the topography to collect in hollows and low lying areas. It is in these low areas that the more severe damage occurs. Lake proximity may reduce this damage to some extent.

Advection freezes commonly occur when large, cool air masses from the mid continent move across the Florida peninsula. The cold fronts travel quickly across the state, accompanied by strong winds. Slight temperature differences occur between low and high lying areas due to wind mixing. Most of the accompanying vegetation damage occurs in hilltop areas, unsheltered from the wind. Advective freezes seldom last more than 2 or 3 days; during day light hours the temperature normally climbs well above the freezing point.

Northern Florida experiences larger annual temperature extremes than does southern Florida because the continental influences are greater over the northern part of the state. Cold spells, caused by cold polar air, usually moderate somewhat by the time the polar front has reached the latitude of central Florida.

JET STREAMS

Global circulation of air is not limited to the lower levels of the troposphere. There is also an exchange of air in the upper levels of the troposphere which is necessary to keep a balance in the global circulation. However, the flow at the upper level follows the parallels of latitude in easterly and westerly directions as contrasted to the northerly and southerly movement of surface air.

Air rising over the ITCZ cools and is deflected to the east by the Coriolis effect as it drifts poleward from the vicinity of the equator. The combination of pressure gradient and Coriolis effect produces upper-troposphere winds of high velocity in the region of 30° North and 30° South latitudes. Frequently these subtropical jet streams reach velocities exceeding 145 knots (167 mph).

In the middle latitudes, high altitude jet streams also occur. They are usually found during the winter just equatorward of the largest temperature contrast, which is usually located between latitudes 35° and 45°, but they may occur as far equatorward as 30° latitude. The summer location of the polar jet stream is about 10° further poleward than the average wintertime position. Day-to-day variations may change the location of the polar jet stream (See Figure 3-6).

The jet stream may be visualized as a narrow stream of high velocity air frequently broken and often not uniform in width or thickness. Characteristically, a jet stream is thousands of kilometers long, a few hundred kilometers wide and from one to two kilometers thick. They are usually located at elevations of eight to 125 kilometers.

The polar jet stream has the most effect upon the climate of Florida during the Northern Hemisphere winter. Although the path is usually located between 35° to 45° north latitude, several times each winter it may dip as far south as 30° North latitude. If the jet stream dips this far southward, when a polar front is moving eastward, it usually means an extremely cold spell for most parts of Florida. Whether the polar front causes the dip in the polar jet stream or the polar jet stream guides the polar front into Florida is not known at the present time. However, there is a positive correlation between the location of the polar jet stream and the severity of the weather in Florida.

TEMPERATURE INVERSION

Another phenomenon that occurs in Florida, especially during the winter, is temperature inversion. Although temperature usually decreases with increasing elevation, weather conditions in the lower air are sometimes reversed. Instead of decreasing, the temperature first rises with increasing elevation before beginning to drop off into the normal lapse rate. This condition is known as temperature inversion, and signifies that warmer air overlies colder air. It commonly results at night from rapid heat loss by reradiation from the ground surface and bottom air layer upward into spaces. (See Figure 3-7). Calm air and clear skies favor its development; whereas turbulence or unsettled conditions prevent its occurrence. If the heat loss is great during the night, the air temperature close to the ground may drop below freezing, resulting in a killing frost that damages sensitive crops. Of course, this is not the cause of all inversions, much less all frosts. Other causes involve moving air masses and are not phenomena of heat radiation. Killing frosts during inversions are sometimes prevented in citrus groves by initiating a circulation of air that mixes the warmer air above with the colder air layer near the ground. This may be accomplished by the use of heaters which set up air currents or by powerful motor-driven propellers. Both methods are used in Florida.

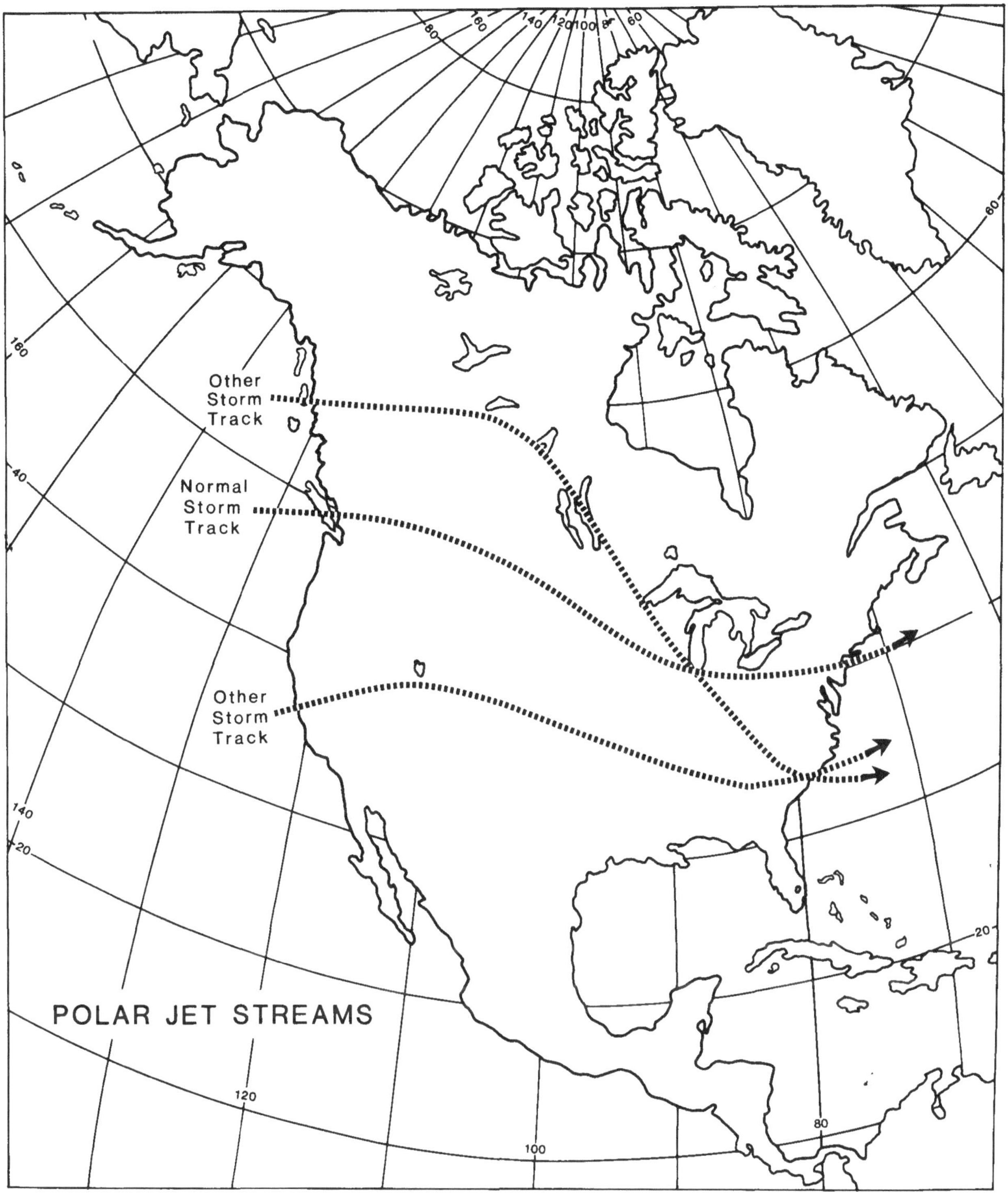

Figure 3-6. The Polar Jet Stream has the greatest effect upon the climate of Florida during the winter months when it may dip as far south as 30°N and bring cold continental air into the state. The Polar Jet Stream may change its path from day to day. Three of the many possible paths are shown on the above map.

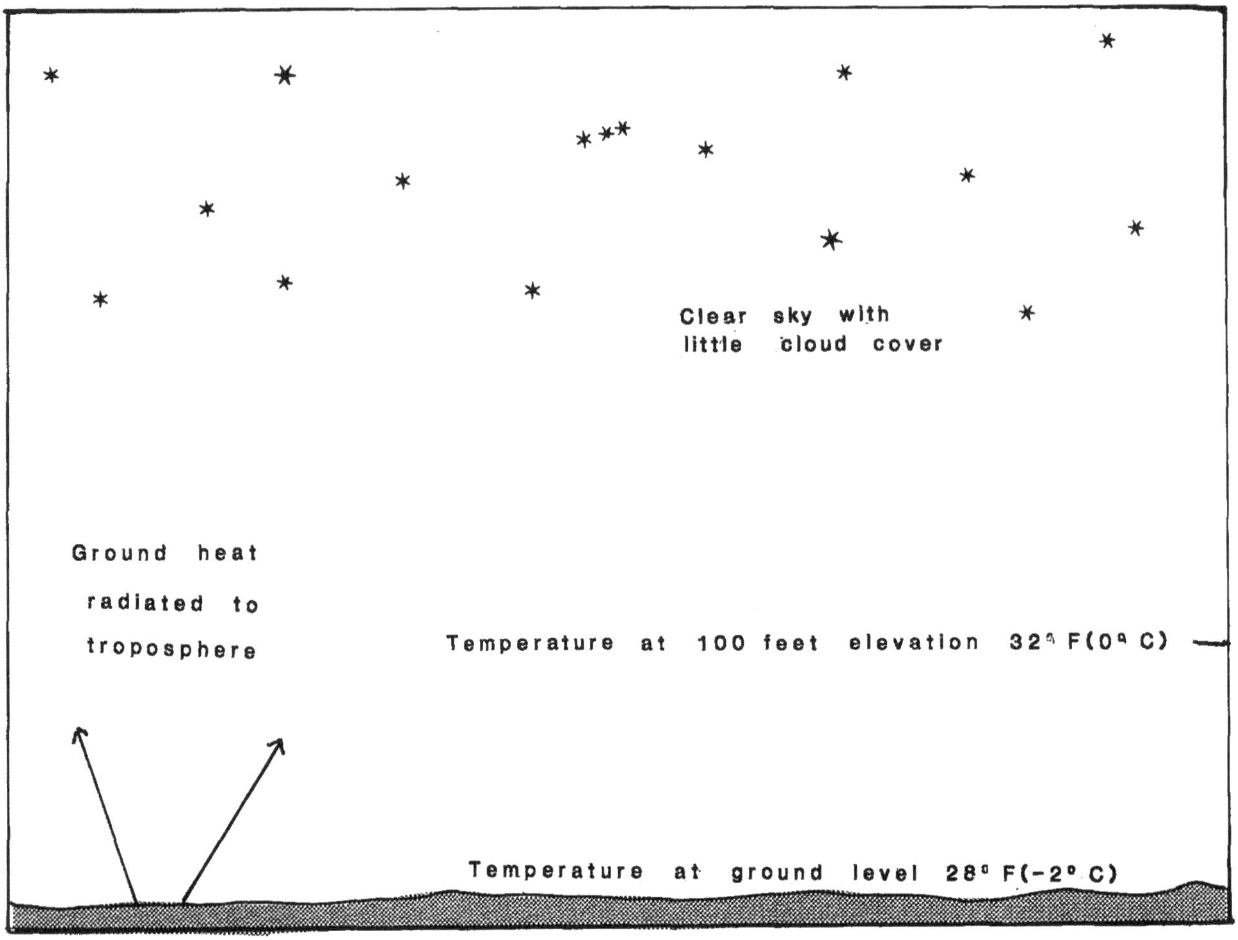

Figure 3-7. Temperature inversion. On cold, clear winter nights the earth radiates its heat upward into the troposphere as soon as the sun sets. This results in the temperature near the ground being colder than the temperature of the air above the ground

PRECIPITATION

Florida receives approximately 60 percent of its annual precipitation during the months of June through September. This rainfall comes from convectional precipitation, the third type of rainfall. The other two types, orographic and frontal, have been discussed earlier in the chapter. Convectional rainfall results from the heating of the earth, which in turn radiates the heat to the adjacent air. The warm air then rises, expands, and cools. If the air rises high enough to cool and reaches the dew point, condensation takes place.* This results in clouds and frequently precipitation.

During the early months of summer, clouds will begin to form shortly after midday as the ground heats up and the heavy black cumulonimbus clouds gather in the sky. By 3 p.m., or shortly thereafter, thunder and lightning may be seen, followed by a heavy downpour of rain. The thunderstorms will occur earlier in the day in July, and still earlier in August, and slacken off in September. The decrease in precipitation is more evident in central Florida and in northwest Florida than it is along the lower southeast coast. At some stations along the southeast coast, such as Miami, there is a second maximum that has been attributed to the hurricane effect; however, Brandes (1982) made a study of 19 stations in central and south Florida during the period 1919-1978 and could not find proof that

* Dew point is the temperature at which a sample of air becomes saturated.

the fall maximum was influenced to any large extent by hurricanes and tropical storms.

As shown in Figure 3-8, two general areas of Florida receive 60 or more inches of rainfall annually. The larger area is in northwest Florida, extending from just west of Panama City to the Alabama border. A much smaller subregion is located west of Tallahassee and south of Quincy. The second general area lies along the lower southeast coast and is divided into two smaller areas: one just south of Palm Beach and the other encompassing the Miami-Homestead District.

Almost one-third of the annual rainfall in Pensacola falls during the months of December through March, while only 13 percent falls during the same months in Miami. Because Pensacola lies closer to the interior of the land mass of the continent, it receives more winter rainfall from polar fronts than the lower southeast coast. By the time the polar fronts reach the Palm Beach-Miami area, they have moderated and produce less precipitation. Northwest Florida receives approximately 51 percent of its rainfall in the form of summer convectional showers during May through September, compared to 62 percent for the same period in Miami. The Gulf Stream lying just offshore of the Palm Beach-Miami coast supplies low level moisture to the southeasterly winds which blow across it toward the shore. Upon reaching the shore, the coastal ridge provides the heat and instability to force the air to rise, cool, and condense, resulting in convectional showers. Less than 10 miles away across Biscayne Bay in Miami Beach, the mean annual rainfall is slightly over 46 inches compared to Miami's 60 inches. This is a good example of the effects of land mass upon precipitation. The lower rainfall for Miami Beach results from the small land area of the narrow bar protruding into Biscayne Bay. The moisture-laden winds crossing the Gulf Stream are not heated sufficiently as they pass over Miami Beach to rise high enough to reach the dew point and produce rain. The island on which the city of Key West is located is also too small an area to heat sufficient quantities of air to rise, reach the dew point, and produce precipitation.

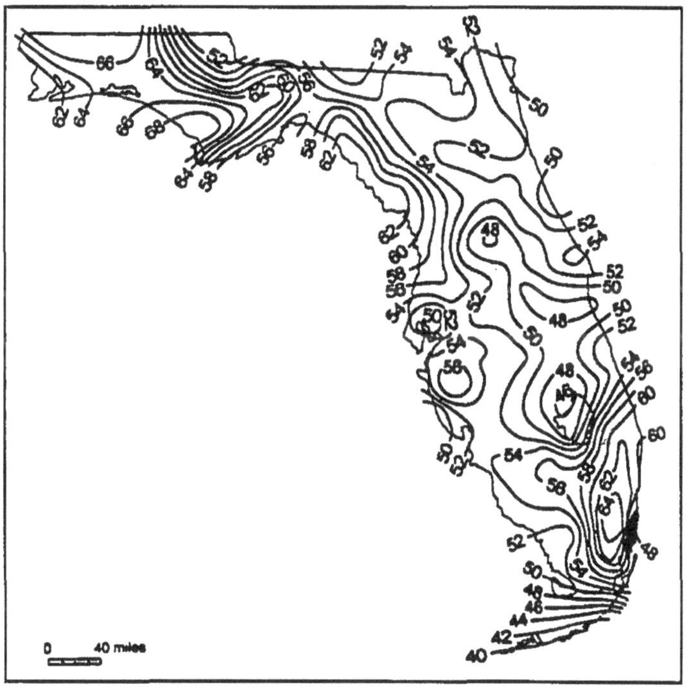

Figure 3-8. Map of annual precipitation. Most of the State of Florida receives the major part of its rainfall during the months June through September. The northwestern part of the state also receives considerable cyclonic precipitation from polar fronts moving into the area during the winter. (Adapted from The Climate and Weather of Florida by James A. Henry, Kenneth M. Portier, and Jan Coyne. Copyright © by James A. Henry, Kenneth M. Portier and Jan Coyne. Reprinted by permission.)

Thunderstorms are a common phenomena over Florida as shown in Figure 3-9. The Tampa Bay region has long been considered the thunderstorm capital of North America. In a 10-year study, Stowers (1978) found that there was reason to believe that the thunderstorms were largely the result of several factors working together: (1) The convergence of daily sea breezes begin shortly after sunrise and reach their peak in mid-afternoon; (2) the rapid development of urbanized areas have created "heat islands," which in turn cause pockets of strong local convergence; (3) the high angle of the sun during the summer months results in the intense heating of the light, sandy soil lying between ridges, which in turn reradiates the heat to the adjacent air, providing an optimum environment for early morning sea breezes; (4) the high humidity during the summer months supplies moisture, keeps the temperature relatively high day and night, and provides a contrast between the warm night air and the cooler ocean air off the east coast. This results in sea breezes beginning several hours earlier on the east coast than they do on the west coast where the Gulf waters are warmer; (5) the orientation of Tampa Bay allows breezes to flow 20 miles from a southwesterly direction with little friction before entering the urbanized land area of the city of Tampa and its environs.

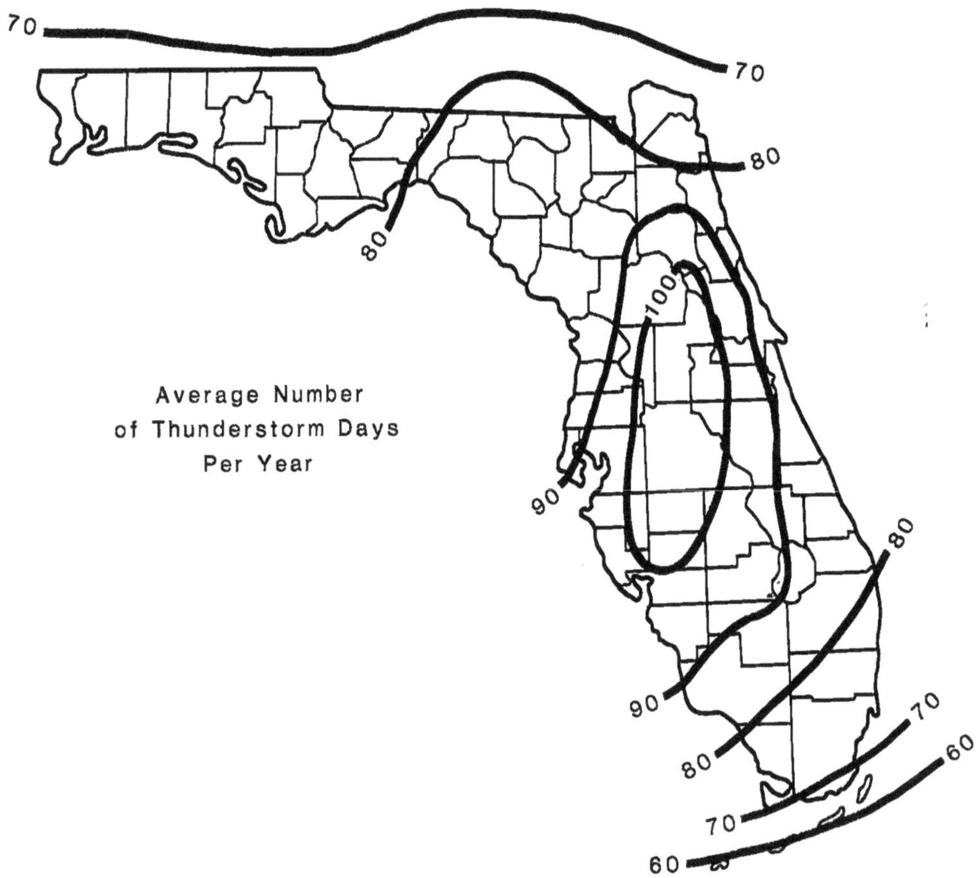

Figure 3-9. The Tampa Bay region is the thunderstorm capital of the continental United States. The isopleths indicate the number of thunderstorms per year. (Reprinted by permission of Dr. Dewey Stowers.)

The combination of the above five factors results in the convergence of the warm, moist air from Tampa Bay with the sea breezes from the east and west coasts at inland locations where the air obtains sufficient heat to rapidly rise, cool, and reach the dew point. This results in thunder, lightning, and heavy showers.

DROUGHT

Florida is not immune to drought, even though annual rainfall totals are relatively large. Prolonged periods of deficient rainfall are occasionally experienced, some for several successive years, even during the summer months. More periods of drought are forecasted for the future because of the over draining of the wetlands, the heavy pumping from the aquifers, and the drying up of many lakes and ponds.

There are several causes of drought which frequently occur at the same time. One of the more important reasons is the subsidence of air from higher to lower altitudes. Subsiding air compresses as it descends towards the earth's surface. The warming of the air that accompanies compression decreases relative humidity, makes the air more stable, and reduces the formation of clouds. These factors result in high pressure areas over the State of Florida.

High pressure systems may also migrate from west to east with the upper wind currents and stagnate over the state. Descending air can originate from high pressure cells in the upper atmosphere over the interior of the United States. The winds from these cells flow from the north toward Florida. Droughts may also result when the Azores-Bermuda high pressure system moves too far west of its normal position and forms a high pressure system over the area just to the north of Florida. Meteorologists are at a loss to explain why the Azores-Bermuda high moves westward and/or northward of its normal location.

Because a large part of the agricultural produce is planted, grown, and marketed during the dry fall, winter, and spring months, many growers have concluded that it is almost mandatory to provide supplemental irrigation for success. However, heavy rains can occur during these seasons, and growers have found it also necessary to provide water removal facilities in addition to irrigation. Statewide droughts during the summer months are rarer than at other seasons of the year. However, at this time the evapotranspiration rate is so high and the soil so porous that a two- or three-week period without rain can appear as a drought. It is not unusual for a drought to occur in one section of the state at the same time another section is having above average precipitation.

Snowfall in Florida is unusual although measurable amounts have fallen in the northern areas at irregular intervals, and a trace of snow has been recorded as far south as Homestead. The greatest recorded snowfall was in February, 1899. Four inches were measured at Lake Butler in Union County, one-half inch at Bartow in Polk County, and a trace as far south as Fort Myers. A little more than half a century later, in February 1958, most of Florida west of the Suwannee River received two to three inches of snow. Areas east of the Suwannee River and north of latitude 30° North received one to two inches on the same night. In January, 1977, Palm Beach had its first snow on record; Tampa received 0.2 inches, and snow fell as far south as Homestead. The snow was accompanied by winds of 40 to 45 miles per hour, making the wind chill factor equivalent to -15°F (-26°C). The temperatures throughout south Florida for January 19, 1977, were the lowest for that date since the winter of 1917-18. Snow was also recorded in February, 1973, as far south as Gainesville. De Funiak Springs had two inches, while several inches fell along the Georgia border paralyzing traffic for an entire weekend.

WIND

The prevailing winds are from the southeast and east over the southern part of the peninsula. Wind directions are more erratic over the remainder of the state, but follow a general northerly pattern in the winter and southerly in the summer. Convectional heating during the summer is responsible for the local wind direction of interior locations, while land and sea breezes characterize the coastal regions. The highest average wind velocities occur in March and April Local winds of high velocity and short duration accompany convectional thunderstorms in summer and frontal storms during the winter.

An average of about 10 to 15 tornadoes, funnel clouds, and waterspouts occur in Florida in a given year. However, it is possible for one severe storm to foster several tornadoes before it dissipates. A severe frontal system originated along the Pacific coast of California at the end of January 1983. Within a week it had crossed the entire United States, entered northeast Florida, and moved southward the length of the peninsula. The storm spawned some 16 tornadoes, extending from Alachua County to the Florida Keys. One tornado did $1.5 million in damages as it crossed Alachua County in a northeast-southeast direction. It destroyed buildings, uprooted large trees, and killed one man. The February 1983 tornado resembled the May 1979, tornado that struck Gainesville; both began just before dawn and caught people asleep in their homes.

The tornado is a small, intense cyclone in which the air is spiraling at a tremendous velocity. It appears as a dark funnel cloud, hanging from a large cumulonimbus cloud. At its lower end, the funnel may be 300 to 1,500 feet in diameter. The funnel appears dark because of the density of the condensing moisture, dust, and debris swept up by the revolving wind. Wind velocities in a tornado exceed the velocities in other cyclonic storms. Estimates of the wind velocity run as high as 500 miles per hour. There is, in addition, a violent updraft in the funnel. Tornado destruction occurs both from the high wind velocity and from the sudden reduction in air pressure in the vortex of the cyclonic spiral. Closed houses literally explode. Where polar air meets warm, moist tropical air along a cold front, conditions may be favorable for tornadoes. They occur most frequently in the Mississippi Valley and are rare over mountains and forested regions. Waterspouts are similar in structure to tornadoes but form at sea. They are smaller and less powerful than tornadoes. Occasionally waterspouts move inland but affect only small areas because they dissipate rapidly after reaching land.

HURRICANES

In the section on General Atmospheric Circulation the Intertropical Convergence Zone was discussed. The ITCA lies close to the equator where the Coriolis force is weak and there is a reduced tendency toward cyclonic and anticyclonic air flow. In addition, there exists a general low-level inversion layer known as the trade wind inversion. This results from subsidence of relatively warm high-altitude air that forms the subtropical high pressure cells of the horse latitudes. The air close to the surface of the earth in the vicinity of the trade winds is relatively unstable and humid. Thus, the lower moist air is effectively separated from the upper dry air and is characterized by stable conditions. However, the stable conditions of the trade wind region are frequently penetrated by weak disturbances called easterly waves in the Caribbean or equatorial waves in the Pacific. Easterly waves appear as wave-like horizontal wind flow patterns, which usually form in the convergent flow of trade winds and move slowly in a direction from east to west. Each wave has a length of approximately several thousand kilometers and an amplitude of a few degrees of latitude.

Easterly waves provide a mechanism for releasing the stability normally maintained by trade wind inversion. The lifting of the inversion by rising air to the east of the trough can eradicate the inversion, changing the area from one of conditional instability to one of active instability. Thus, thunderstorms can develop, and frequently squall lines accompany the storm.

Some easterly waves develop vortices and become tropical cyclones while others reach hurricane intensity. The violent and destructive cyclones are known as hurricanes in Atlantic regions, Caribbean Sea, Gulf of

Mexico, parts of Australia, and the eastern Pacific. Other names for the same phenomena are typhoons in the western Pacific, cyclones in the Indian Ocean and willy-willies off northwest Australia.

Hurricanes most frequently develop from a low pressure distribution, or easterly wave, within the belt of the trade winds. This belt extends from about 30°N, or 30°S, to approximately 5° of the equator. The necessary conditions to produce an easterly wave or low pressure area, are prevalent during summer and fall. If the Azores-Bermuda high pressure cell is especially weak and shifted to the south of its normal position, a low pressure cell may be introduced into the trade winds belt. Southern sections of such systems become captured in the trade winds providing seeds for a hurricane. The trapped easterly wave then may move from west to east producing clouds and heavy rains.

During the hurricane season, which officially lasts from June 1 to November 1 in Florida, a large majority of the easterly waves have their origins in the Sahara Desert of Africa. Another breeding ground for easterly waves is in the vicinity of Cape Verde Island, east of the coast of West Africa. A classic example of a Cape Verde hurricane occurred in 1988 when hurricane Gilbert struck Caribbean islands and the Yucatan Peninsula before passing through the Keys on its approach to Florida. The barometric pressure was later determined to be lower than the famous Labor Day hurricane of 1928. Calculations estimated a wind speed in excess of 200 miles per hour.

One notable storm originated in the western Sahara Desert in August 1985. It began as a long, narrow low pressure area over the desert, moved eastward over the Atlantic Ocean and within two weeks it appeared in the Gulf of Mexico and followed the west coast of Florida until it stalled between Tampa Bay and Cedar Key. The storm was named Elena. It started again and moved northwest off the coast of the Florida Panhandle and finally turned inland in the vicinity of the Mississippi-Louisiana border. While in Florida, some one and one-half million Floridians were evacuated—the most in the state's history.

Of the hurricanes recorded and tracked over the Atlantic tropical basin about three to four per cent of the major hurricane force struck or affected Florida (rank of three or higher on the Saffir-Simpson Hurricane Scale).

Tropical cyclones are defined as non-frontal, low pressure, large-scale weather systems that develop over tropical or subtropical waters and have a definite organized circulation. The term tropical depression is applied if the sustained winds near the center of the system are at least 33 knots (38 mph). When the sustained winds reach between 34 knots and 63 knots (39 mph and 73 mph) the term tropical storm is applied. Once the winds exceed 64 knots (74 mph) the storm is called a hurricane.

The typical hurricane, as shown schematically in Figure 3-10, has a diameter that can vary from about 95 miles (150 km) to 630 miles (1,100 km); yet diameters as small as 19 miles (30 km) have been observed. Neither the intensity of the storm nor the rate of travel appear to be related to the size of the storm. However, the famous Matecumbe Key (Labor Day) Hurricane of 1935 had a very small eye and held the record for the lowest barometric pressure recorded in the Western Hemisphere, 26.35 inches of mercury (892.35 mb), until 1988 when Hurricane Gilbert had a central pressure lower than the Labor Day hurricane.

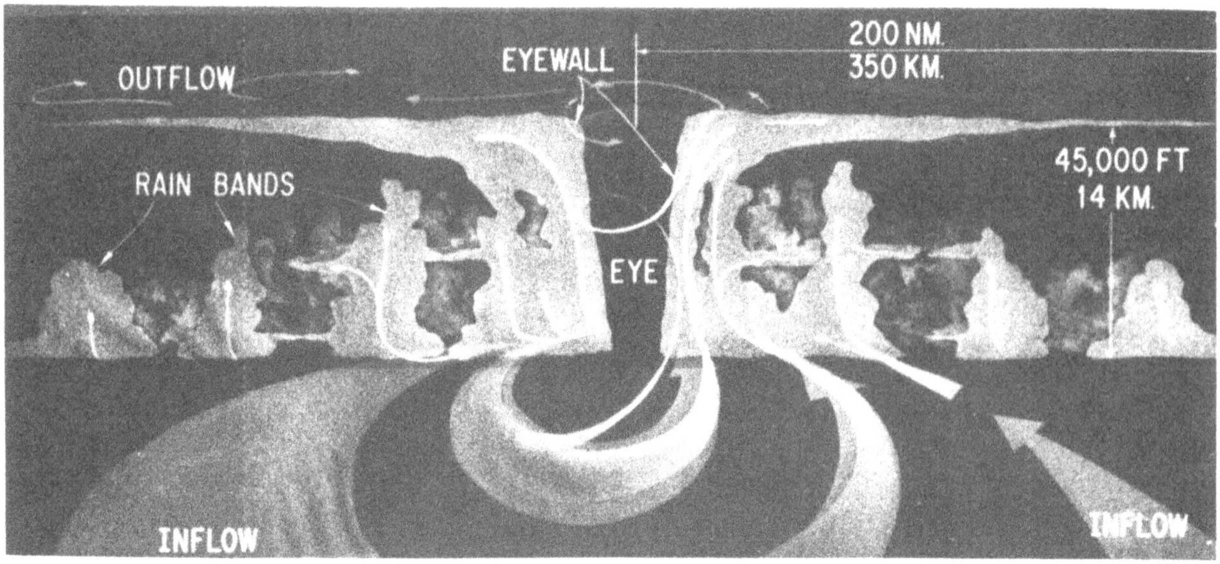

Figure 3-10. Schematic cross-section of a hurricane showing the eye, eye wall, and the rain bands. (Reprinted by permission of National Hurricane Center, NOAA.)

The most characteristic feature of a hurricane is the velocity and the force of its winds. The winds at the outer edge are light and gusty. Moving in toward the center of the storm, the winds increase in intensity with deluges of rain and low-flying clouds. At the center of the storm lies the vortex or eye. Here the wind is calm, the rain ceases, and frequently the sun can be seen shining from above. The swirling mass of rapidly rising air rotates in a counterclockwise direction around the eye in the Northern Hemisphere and in a clockwise direction in the Southern Hemisphere. The storm receives its energy from the condensation of the rapidly rising moist air. This is verified by increases in intensity over water and the dissipation of the hurricane over land.

While a hurricane follows its twisted path, it is studied continuously. When far out at sea satellite imagery such as Figure 3-11 provides important data. As the storms reach closer to shore, monitoring by aircraft and aircraft dropped equipment are also employed. Parachute dropped equipment has been developed to monitor the atmospheric currents that actually steer the storm. The data is used to create a picture of the environment around a hurricane, pinpoint wind direction, and to determine the future course of the storm.

Figure 3-11. Satellite image of Hurricane David at 6:00 PM., September 3, 1979. At the time of the picture, David was located just north of Palm Beach. David was the most intense hurricane of the 1979 season. Before reaching Florida, it was responsible for 1,262 deaths in the Caribbean area, and after hitting the United States mainland five additional deaths were recorded. (Reprinted by permission of National Hurricane Center, NOAA.)

Saffir/Simpson Hurricane Classification Scale

Category	Central Barometric Pressure	Winds (mph)	Storm Surge (in feet)	Damage
1	>28.94	74.95	4-5	Minimal
2	28.50-28.93	96-100	6-8	Moderate
3	27.91-28.49	111-130	9-12	Extensive
4	27.17-27.90	131-155	13-18	Extreme
5	<27.17	155>	18>	Catastrophic

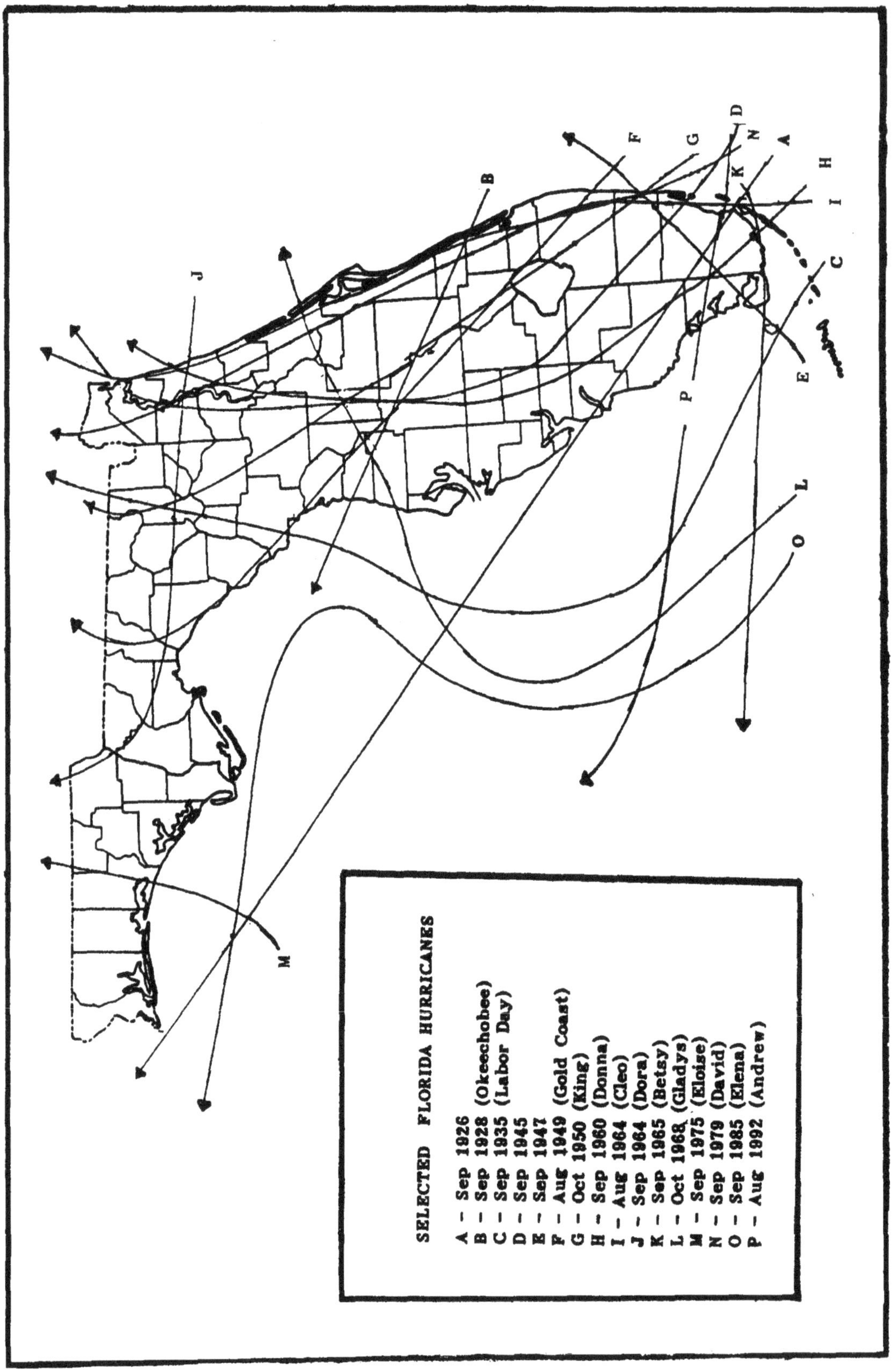
SELECTED FLORIDA HURRICANES
A – Sep 1926
B – Sep 1928 (Okeechobee)
C – Sep 1935 (Labor Day)
D – Sep 1945
E – Sep 1947
F – Aug 1949 (Gold Coast)
G – Oct 1950 (King)
H – Sep 1960 (Donna)
I – Aug 1964 (Cleo)
J – Sep 1964 (Dora)
K – Sep 1965 (Betsy)
L – Oct 1968 (Gladys)
M – Sep 1975 (Eloise)
N – Sep 1979 (David)
O – Sep 1985 (Elena)
P – Aug 1992 (Andrew)

Figure 3-12.

Before 1950, there was no formal nomenclature for identifying hurricanes. They were usually described in terms that would identify the area over which the most damage was recorded, the largest loss of life sustained, the lowest barometric pressure recorded, and the date when the hurricane struck. An excellent example of such nomenclature is the Matecumbe Key Hurricane. It is most frequently referred to as the Matecumbe Key Hurricane or the Labor Day Hurricane of 1935.

In 1950, an official nomenclature for tropical cyclones began. At first the phonetic alphabet was used to name the hurricanes in chronological sequence. Examples of the first five letters of the phonetic alphabet are Able, Baker, Charlie, Dog, Echo. This nomenclature was not entirely satisfactory, so in 1953 the practice of using women's names was adopted and continued until 1977 when the "sexism" was deleted and both men's and women's names became official.

CULTURAL ASPECTS OF HURRICANES

There are four destructive aspects of hurricanes: (1) high winds, (2) heavy rainfall, (3) high tides, and (4) tornadoes. Florida has felt the effects of all of these.

The Matecumbe Key Hurricane had a wind velocity of 174 knots (200 mph) and a barometric pressure of 26.35 inches, the lowest on record for North America. Florida's flat terrain frequently magnifies the flooding that accompanies a hurricane, such as the 16.40 inches that fell within a few hours near Blountstown in the 1926 Miami hurricane or the 38.7 inches that fell at Yankeetown from Hurricane Easy in 24 hours during the period September 3 to 7, in 1950. Frequently areas have had heavy rain before the arrival of the hurricane, and this contributes to severe flooding.

Hurricanes are unusually disastrous in September because they occur during a period when tides are higher than usual and especially when they bring with them the dreaded tidal wave. Cedar Key was swept away twice by hurricane waves: first in 1842 when a 27-foot wall of water rolled over what was then a military outpost and again in 1896. A study of hurricanes reveals that wave (storm surge) or water damage from hurricanes is often more destructive than wind damage and accounts for a much larger toll of life.

The last destructive force that results from hurricanes is from the tornadoes that sometimes accompany them. These are infinitely smaller in area, but the wind velocity frequently exceeds the hurricane that spawns them, and area for area matches the hurricane in destructiveness. Fortunately, they are not too frequent.

Despite sophisticated warning devices that are available with modern technology, weather experts and environmentalists are deeply concerned about the effects of a severe hurricane along the coasts of Florida, especially the southwest coast. In June, 1982, a storm with wind gusts of only 60 miles per hour did $50 million in damage along the southwest coast from Pinellas County to Collier County.

Since the time that Donna passed through Florida in 1960 with winds of 113 knots (130 mph), the population of southwest Florida has tripled, the number of coastal high-rises and condominiums has greatly increased, and the escape routes have not been built to keep pace with the increase in population. In addition, many of the newcomers to this area have never experienced a hurricane. Without sufficient warning the inhabitants of Sanibel and Captiva Islands may not have time to evacuate if another hurricane of Donna's intensity follows the same or similar path. It has also been estimated that a storm of Donna's magnitude could inflict property damage as high as $4.5 billion with the added possibility of a large loss of life. Typical hurricane damage is shown in Figures 3-13 and 3-14.

Figure 3-13. Damage to a mobile home by Hurricane Betsy in 1965. Mobile homes throughout the state are particularly vulnerable in hurricanes, and this is typical of their destruction. (Reprinted by permission of National Hurricane Center, NOAA.)

Figure 3-14. Mangrove killed by Hurricane Donna in 1960 near Flamingo in Everglades National Park. Hurricanes do not respect the natural or cultural environment.

The destructiveness of a Category 4 hurricane was further substantiated by Hurricane Andrew on August 24, 1992 when it hit Homestead and surrounding areas at 5:00 a.m. in the morning with sustained winds of 140 mph and gusts exceeding 200 miles per hour. Despite having a storm surge of only five to eight feet, Andrew destroyed more that 25,000 homes and damaged an additional 101,000 homes. It was the most expensive natural disaster in the history of the United States, inflicting damages of $20 to $30 billion. Five years after the hurricane struck and the Federal Government's donation of $823 million in grants to help build the area's roads, utilities, businesses, and homes; more than 300 families have not had any rebuilding done. Some of this results from lack of insurance, fraudulent work, and vandalism.

The damage by Andrew would have been much greater had the storm surge been typical of a Category 4 hurricane, 13-18 feet, rather than the actual five to eight feet. Also, the storm followed almost a direct east to west path across one of the narrowest parts of the peninsular Florida at a velocity of 20 miles per hour ground speed. The storm's trip across Florida took just about four hours.

Hurricanes have played havoc in Florida throughout its history. The first recorded hurricane destroyed seven of 13 vessels of the fleet of Don Tristan de Luna y Arellano in Pensacola Bay on September 19, 1559. This hurricane delayed Spanish settlements in west Florida for almost a century. While the 1559 hurricane delayed Spanish settlements in west Florida, the September 1565, hurricane destroyed much of the French Huguenot fleet and Fort Caroline and speeded up the founding of St. Augustine by Menendez.

Again in 1754, a hurricane struck Pensacola and destroyed the governor's house, a church, and several other buildings on Santa Rosa Island, drowned a number of its inhabitants, and thwarted a second attempt by the Spanish to colonize west Florida.

Hurricanes also took their toll of gold-laden Spanish galleons during their annual voyage from Mexico through the Straits of Florida to Spain. One of the more costly hurricanes occurred in 1715, when several vessels were sunk off present-day Fort Pierce. Losses in Spanish gold and ships by hurricanes have been estimated at over one billion dollars.

Tropical storms have also influenced man's activities since the United States purchased Florida in 1821. Repeated storms in the 1840s are thought to be partly responsible for cattlemen in the Apalachicola Valley moving south and establishing the cattle industry on the grasslands around Kissimmee.

A few more recent hurricanes are listed in Table 3-3. As serious as hurricanes are, some good has come from them. The flimsy cardboard buildings erected in Miami during the boom days were blown away by the 1926 hurricane and replaced by substantial buildings. A strict building code was also enacted to ensure the safety of the inhabitants in this area. Successive flooding by hurricanes in the Lake Okeechobee area helped bring about the establishment of the Central and South Florida Flood Control District, a federal agency which built dikes around the lake and established a more satisfactory drainage system. The overseas railroad in the Florida Keys, damaged during the Matecumbe hurricane of 1935, has been supplanted by the beautiful overseas highway. Lastly, hurricanes frequently bring much-needed water to south Florida during periods of low rainfall. Man cannot prevent hurricanes although he is attempting to modify them through cloud seeding. But he can learn to live with them and through technology give adequate warning in advance. This affords Floridians the opportunity to board up buildings and evacuate low areas.

Table 3-3. Selected Hurricanes That Did Damage in Florida

Hurricane*	Date	Central Pressure (inches)	Saffir/Simpson Damage Potential	Lives Lost	General Location
	Sep 1926	27.61	4	243	Miami, Lake Okeechobee
San Felipe	Sep 1928	27.43	4	1,836	Lake Okeechobee area
"Labor Day"	Sep 1935	26.35**	5	408	Matecumbe Key
	Sep 1945	28.09	3	4	Homestead and other parts of Dade County
	Sep 1947	27.97	4	17	Hillsboro Light, Fort Lauderdale
Gold Coast	Aug 1949	28.17	3	2	Center passed over Lake Okeechobee
King	Oct 1950	28.20	3	6	Miami area
Donna	Sep 1960	27.98	4	13	Marathon Key and extreme so. Florida
Cleo	Aug 1964	28.58	2	11	Miami northward along east coast to Jacksonville
Dora	Sep 1964	28.57	2	1	Jacksonville and St. Augustine
Betsy	Sep 1965	28.12	3	13	Florida Keys and Lower East Coast
Gladys	Oct 1968	28.50	3	3	Between Bayport and Crystal River
Eloise	Sep 1975	28.20	3	4	Storm surge from Ft. Walton Beach to Panama City
David	Sep 1979	27.55	4	<5	Along East Coast
Elena	Sep 1985	28.08	3	3	55 miles west of Cedar Key
Andrew	Aug 1992	27.23	4	17	Homestead area

*Hurricanes were not officially named until 1950

**This was the lowest pressure ever recorded in the Western Hemisphere until Gilbert struck in 1988 and produced the lowest barometric pressure recorded in the Atlantic Ocean, Caribbean Sea, or the Gulf of Mexico. (Adapted from Henry, Portier, and Coyne, 1994.)

Table 3-4. Chances of Hurricane Force Winds in Any Given Year

City	Chances	City	Chances
Jacksonville	1 in 100	Fort Myers	1 in 11
Daytona Beach	1 in 50	Key West	1 in 8
Tampa-St. Petersburg	1 in 25	Pensacola	1 in 8
Melbourne-Vero Beach	1 in 20	Palm Beach	1 in 7
Apalachicola-St. Marks	1 in 27	Miami	1 in 6

(Adapted from Henry, Portier, and Coyne, 1994.)

El Niño and La Niña

El Niño is a massive warming of the ocean-atmosphere system in the eastern Pacific. Its cause is unknown. El Niño occurs at approximately seven-year intervals and is characterized by dwindling or reversal of the trade winds which normally blow from east to west. In Summer and Fall easterly lows (including hurricanes) either do not form in the Atlantic or do not drift westward to drench Florida. Thus our wet season changes to one of drought. El Niño also shifts the jet stream (Figure 3-6) to a more southerly course, resulting in more Florida tornadoes, and making normally dry winters wetter and cooler.

La Niña is a reversal of its brotherly phenomenon, with colder than normal temperatures in the Pacific and warmer waters in the Atlantic. This results in global weather changes opposite to those of El Niño, with extensive rainfall and hurricane activity in Florida.

These two events each last about a year to eighteen months. We now can determine when they are about to begin and their stage of development by use of satellites, a series of sea bouy systems, and other monitoring techniques.

REFERENCES

Brandes, D. (1982), "Hurricane Rainfall in Florida," *The Florida Geographer*, Vol. 16, No. 1, pp. 17-24.

———. (1981), *The Significance of Tropical Cyclone Rainfall in the Water Supply of South Florida*, unpublished Ph.D. dissertation, University of Florida, Gainesville, Florida, 150 pp.

Critchfield, J. H. (1983), *General Climatology*, Prentice-Hall, Inc., Englewood Cliffs, New Jersey, 453 pp.

Dunn, G. E. and B. I. Miller (1960), *Atlantic Hurricanes*, Louisiana State University Press, Baton Rouge, Louisiana, 326 pp.

HO, F. P., R. W. Schwerdt and H. V. Goodyear (1975), "Some Climatological Characteristics of Hurricanes and Tropical Storms, Gulf and East Coasts of United States," NOAA, National Weather Service *Technical Report*, 87 pp.

Henry, J.A., Portier, K.M. and Coyne, J., *The Climate and Weather of Florida (1994)*, Pineapple Press, Sarasota, Florida, 279 pp.

Ludlum, D. M. (1963), *Early American Hurricanes, 1942-1970*, American Meteorological Society, Boston, Massachusetts, 198 pp.

Marcus, R. B. and E.A. Fernald (1975), *Florida: A Geographical Approach*, Kendall/Hunt Publishing Co., Dubuque Iowa, 302 pp.

Marcus, R. B. (1964), *A Geography of Florida*, W. C. Brown Company Publishers, Dubuque, Iowa, 159 pp.

Miller, B. I. (1964), "A Study of the Filling of Hurricane Donna (1960) over Land," *Monthly Weather Review*, Vol. 92, No. 9, pp. 398-406.

———. (1967), "Characteristics of Hurricanes," *Science*, Vol. 157, No. 3795, pp. 1388-1399.

Myers, R. L. (1986), "Florida's Freezes: An Analog of Short-Duration Nuclear Winter Events in the Tropics," *Florida Scientist*, Vol, 49, No. 2, pp. 104-115.

NOAA (1978), "Climate of Florida," *Climatography of the United States*, No. 60, National Climatic Center, Asheville, North Carolina, 28 pp.

Neuman, C. J., G. W. Cry, E. L. Caso and B. R. Jarvinen (1978), *Tropical Cyclones of the North Atlantic Ocean, 1871-1977*, National Climatic Center, Asheville, North Carolina, 178 pp.

Saffir, H. S. (1977), "Design and Construction Requirements for Hurricane Resistant Construction," American Society of Engineers, New York, *Preprint* Number 2380, 20 pp.

Shapiro, L. J. (1977), "Tropical Storm Formation from Easterly Waves: A Criterion for Development," *Journal of Atmospheric Science*, Vol, 34, No. 7, pp. 1007-1021.

Simpson, R. H., N. L. Frank, D. Shideler and H. M. Johnson (1967), "Atlantic Tropical Disturbances," *Monthly Weather Review*, Vol. 96, No. 4, pp. 251-259.

Stowers, D. M. (1978), "The Tampa Bay Area, The Thunderstorm Capital of North America: A Climatological Analysis," paper delivered at the 74th Annual Meeting of the Association of American Geographers, New Orleans, Louisiana, 27 pp.

Stowers, D. M. and M. S. Le Vaseur (1981), "The Freeze of January 13, 1981: An Impact Study of West Central Florida," paper delivered at the Annual Meeting of the Southeastern Division of the Association of American Geographers, Atlanta, Georgia, 20 pp.

———. (1980), "The Tampa Bay Storm of May 8, 1979," *The Florida Geographer*, Vol, 14, No. 2, pp. 15-20.

Tannehill, I. R. (1956), *Hurricanes*, Princeton University Press, Princeton, New Jersey, 308 pp.

Winsberg, M. D., 1990, *Florida Weather*, University of Central Florida Press, Orlando, Florida, 171 pp.

SUMMARY OF CHAPTER 3

The (1) location, size and shape; (2) topography; (3) planetary wind and pressure systems; and (4) the warm Gulf Stream are all factors that contribute to the type of climate that characterizes Florida.

There are three types of rainfall found throughout the world: orographic, frontal, and convectional. Only convectional and frontal types are found in Florida. Approximately 60 per cent of the total rainfall for the state comes during the summer months by convectional rainfall, while frontal precipitation falls during the winter months when polar fronts move into the area.

Florida is located at the southeastern tip of the North American continent, which places it in or near the subtropical high, which in turn is the source of the northeast trade winds and the prevailing westerlies. This location results in winters being relatively mild and the summers cooler than inland locations of the same latitude.

Hurricanes are an integral part of the climatic pattern of Florida and can be expected during any month between June and November. However, of 500 hurricanes that have tracked and recorded over the Atlantic tropical basin during the period 1886-1992, only 17% had an impact on Florida.

REVIEW OF CHAPTER 3

Part A—Map Exercise:

Color the average annual precipitation map in the appendix, using an appropriate color coding system.

Part B

1. All factors being equal, all areas having the same longitude receive the same amount of solar energy.

   True/False

2. Florida's long peninsula shape is of significant importance in creating a climate of hot summers and freezing winter.

   True/False

3. Approximately one-third of Florida's precipitation is due to orographic causes.

   True/False

4. Florida's winds are due primarily to

   a) Pressure differences between two air masses
   b) falling air within the same air mass
   c) rising air within the same air mass
   d) velocity changes
   e) both a and b
   f) both a and c
   g) all of the above

5. In the Intertropical Convergence Zone (ITCZ) we normally find

   a) a high pressure zone due to rising air
   b) a low pressure zone due to rising air
   c) a high pressure zone due to the cooling effect of rising air
   d) a high pressure zone due to the cooling effect of falling air

6. Florida's climatic pattern shows a monsoonal effect resulting from

   a) southeast summer winds and northwest winter winds
   b) prevailing winds from the Gulf and Atlantic
   c) nightly breezes from the land and sea breezes during the day
   d) summer continental winds and winter sea winds

7. In a category 4 hurricane you would expect wind speeds of _________ and a storm surge of about _____________ feet

   a) 120 mph; 10 feet
   b) 165 mph; 22 feet
   c) 100 mph; 16 feet
   d) 140 mph; 15 feet

8. The path followed by a hurricane striking the Florida coast is affected primarily by
   a) local atmospheric currents
   b) ITCZ winds of the Southern Hemisphere
   c) Bermuda High
   d) winds and storms in the vicinity of the Cape Verde Islands

Part C

1. Why is there such a difference in the average annual rainfall between Miami and Miami Beach? Between Miami and Key West?

2. What is meant by climatic controls? Which one of the climatic controls has the least effect on Florida?

3. What part of Florida receives the largest amount of rainfall? At what season of the year does this occur? Why does Pensacola receive more winter rainfall than Miami?

4. What is the difference between
   a) a high and a low
   b) a cyclone and a tornado
   c) a cyclone and a hurricane

5. What is meant by a "cold spell" in Florida? How does a "cold spell" differ from a severe freeze? What are the effects of a severe freeze on citrus and fishing?

6. Distinguish between the two freezing mechanisms experienced in Central Florida.

7. How do Florida citrus growers use the warm air inversion to protect their crops?

8. List the important factors which contribute to the formation of a hurricane. What is the hurricane season in Florida?

CHAPTER 4
The Face of Florida Today

We have seen the work of eons in creating the peninsula of Florida. Its formation has been studied in a broad geologic brush stroke across the intervening ages. Now let us examine the result of nature's handiwork—"the face of Florida today."

We should recognize that Florida is not completely unique and that its forms and features do not stop at the state political boundaries. The Floridan peninsula is merely a continuation of the coastal lowlands of Georgia and Alabama to the north. As such, the terrain of the plateau is nearly level but dips seaward along its flanks into dunes and beaches.

Florida's interior hills are gently rolling and smooth. The form of these hills is dictated by the physical characteristics (e.g., repose angle) of the sands which constitute them. The sand hills reach their highest elevation near Lakewood (Walton County) where they climb to some 345 feet above sea level. Surface-exposed Ocala limestones are eroded and dissolved by the rains until they are rough and pockmarked. Only minimal limestone outcrops exist; these are seen particularly along the west coast. Much of the Florida interior countryside lies at elevations between 50 to 100 feet above mean sea level.

The topography of the peninsula is strongly influenced by four primary natural forces. These include (1) surface erosion by running water; (2) solution action by groundwater; (3) the work of winds, waves, and ocean currents; and (4) sea level fluctuations. Man's activities are becoming progressively important topographic influences as well.

Running water as an erosion force is most effective in the Panhandle and north central portions of the state. These areas contain soils with a relatively higher clay content than found elsewhere along the peninsular surface. The small clay particles tend to fill the voids between the larger soil grains and to seal these natural openings against infiltrating rain waters. This reduced infiltration results in larger volumes of overland flow during storms. Overland flow exerts a dragging force over the soil surface. It is quite capable of moving a variety of soil size particles. Typical erosional movement varies from picking up and transporting small grains to down slope rolling of gravel and larger stones. It has been estimated that this work is accomplished by less than 10 per cent of the energy of the overland flow. The remainder of the energies are expended in moving through the ground cover and in friction. Additionally, the overland flow is capable of dissolving mineral matter from the Florida soils. The most soluble soil constituents include carbonates and silica, particularly when the flow waters are acidic. The eroding capacity of overland flow is directly proportional to the rate of rainfall, slope length, and slope steepness, and is inversely proportional to soil infiltration abilities and vegetational cover. Erosion by running water has less effect throughout the coastal, central, and southern portions of Florida. In these areas relatively thick sand deposits blanket the surface. While sand movement requires less erosional energies than do clays (clays are cohesive and tend to bond or aggregate together), sandy soils tend to remain in place, as sands encourage rapid percolation of rainfall and surface waters. Thus, drainage throughout most of the state is underground rather than as overland flow. Infiltrating waters pass down through the sands and eventually into the underlying limestone layers. As these waters percolate down they tend to carry the smaller clay particles to lower subsurface levels. The clay removal effectively lowers surface resistance to infiltration. As subsurface drainage increases, surface streams become smaller and fewer in number. This is particularly true in southern central Florida where sand deposits are over one thousand feet thick.

Studies of drainage basins and stream loads have been conducted to determine the average reduction rate of land surface elevations. The measurements contain many inaccuracies but probably yield results of a proper magnitude. The studies indicate an average annual soil loss of approximately 77.2 tons per square mile and an elevation reduction rate of about 1.2 inches per 1,000 years. This erosion rate is enough to reduce Florida to sea level in 3.5 million years if land uplifting forces were not also operative.

The terrain of northern and central Florida is influenced primarily by the action of groundwater. Florida's limestone is strongly resistant to mechanical erosion, but is subject to rapid solution. Dissolution of the limestone is promoted by the presence of carbonic acid in the water. Weak carbonic acids are formed when rainwater dissolves carbon dioxide from the atmosphere and decaying plant matter. Circulating acidic groundwater flows readily into joints, fractures, and between limestone bedding planes. The solution action progresses rapidly along such avenues and produces a system of caverns, tubes, and channels, which enlarge with time. Large underground streams develop through many cavern systems and transport significant qualities of dissolved limestones. Silver Springs near Ocala is part of such a system and discharges approximately 500 million gallons per day. This water contains some 600 tons of dissolved carbonates.

Subsidence and collapse of the overburden into the limestone cavities is a common process. The result varies from small dish-shaped depressions on the surface to deep, funnel-shaped cavities extending down perhaps 100 feet or more. The latter feature is termed a **sinkhole**. The mechanisms relating to subsidence and sinkholes will be discussed in a following chapter. Water-filled sinkholes have produced much of Florida's interior lake system.

Landscapes which occur in areas where limestone solution is particularly active are termed **karst topography**. Such regions are named after a massive limestone area in northwest Yugoslavia where solution sculpturing is well-documented. Karst areas contain: (1) abundant sinkholes (either dry or Water-filled), (2) well-developed cavern systems, (3) underground diversion of surface streams, (4) lapies or deeply grooved and fluted surface limestones, and (5) **prairies** or flat-floored valleys resulting from large coalesced sinks. In the latter case, a particularly well-known prairie—Payne's Prairie—lies just south of Gainesville. A typical Florida prairie is shown in Figure 4-1 and other karst features in Figure 4-2.

Geologic evidence suggests that limestone solution and lowering of the Florida surface has been most extensive in the area of the Ocala uplift where the early Tertiary limestones are exposed. Humid climate studies suggest the Florida surface may be lowered 1 foot per 10,000 years through this process alone. Evidence of such limestone solution appears in Figure 4-3.

Figure 4-1. View of typical Florida prairie. (Reprinted by permission of Florida Game & Fresh Water Fish Commission.)

Figure 4-2. Karst topography of central Florida. (Reprinted by permission of F. B. Kujawa.)

Figure 4-3. Lapies of solution-eroded limestones exposed along the Alapaha River. This outcropping is about 12 feet high and 8 feet wide. (Reprinted by permission of F. B. Kujawa.)

Florida's older Pleistocene terraces tend to reflect greater degrees of solution activity. Beach ridges of solution-modified terraces often vary up to one hundred feet in crest height. By comparison, younger, unaffected ridges often vary in height by only 10 feet over a distance of 50 feet.

Florida's karst or sinkhole topography has been divided into two geomorphic districts. In the lime-sink district, the Ocala limestone is covered primarily with the relatively impervious Hawthorn formation. The protection offered by this formation promotes relatively shallow sinkholes, few ponors, or nearly vertical sided sinks, and a karst topography somewhat comparable to that found elsewhere in North America. Sinking streams and related phenomena are lacking, however, due to the permeability of the overburden. The lake district represents areas where the Ocala limestones are essentially uncovered, the lakes are deeper, and ponor formations more common. The two geomorphic districts are shown in Figure 4-4.

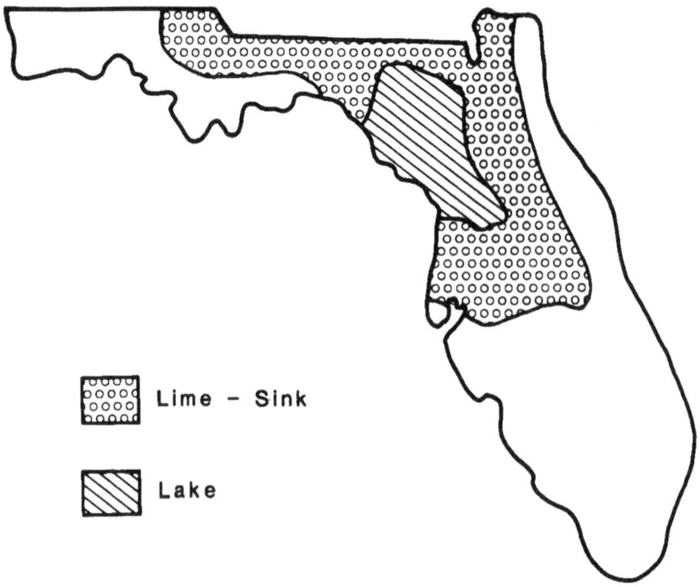

Figure 4-4. Florida karst districts as based on the Hawthorn covering of the Tertiary limestone. In the limesink district the Hawthorn covering is present in variable thicknesses, while in the lake district the limestone is essentially uncovered.

Today's Florida beaches are smooth. The sands make them so. Behind the beaches, gentle mounds or sloping dunes march inland. In many cases old dunes lie miles inland to mark former seashores. Importantly, such dunes should also be recognized as the basis of many ridges existing in the Pleistocene Terraces.

Most of Florida's dunes and ridges began their life in the sea. Their beginnings lie as offshore bars formed by waves breaking in the shallow waters of the "breaker zone." Frictional forces produced by waves moving over the shallow and gently sloping coast scoured loose sands from the ocean bottom. The scoured sands gathered on the breaker zones where they formed submerged ridges or bars. The bars continued to grow until finally they rose above the surface of the waves. Along Florida's east coast shifting longshore currents provided much of the sands and shells for bar building. Large volumes of east coast sands were pulled south from Appalachia by currents counter to the Gulf Stream. The bars grew to such size as to create barrier islands, which were separated from the mainland by shallow lagoons. Similar bars formed and grew in the Gulf, fed by less active Gulf currents.

Nature's forces of wind and waves attacked the seaward flanks of the new bars and islands. Progressively the bars were rolled landward and slowly but inexorably forced to cross the lagoon. Somewhere in the lagoon they joined land-derived lagoonal deposits to create an extended mainland. Continued wave action enlarged the sand deposits and formed ever-widening beaches.

Dried sands on the upper beaches were driven by onshore winds to form a series of offshore dunes. Generally waves deposited new beach sands at a rate somewhat equal to the rate of dune formation. Shore dunes advanced landward with a broad frontal movement where the "wind had mastery" of the coast. Soon, small embryonic dunes merged into wide transverse ridges (often just a single ridge) paralleling the coast. The coastal parallel pattern of the sand dunes and ridges can be attributed to the moderate wind velocities (strong winds produce dunes parallel to wind direction or perpendicular to the coast) and a constant sand supply. The progress of dune frontal line advancement is somewhat obscured by the rapid appearance of vegetation on the emerging beach. Seeds and plants initially gain a foothold in debris heaped in rows by high tidal action. With pioneer plants marshalled in broad lines, a continuous sand supply, and onshore winds, a foredune emerges parallel to the shoreline and grows to its final elevation. Eventually, each foredune position is in turn usurped by the formation of a new foredune seaward of its position.

The windward side of dune ridges maintain a smooth, gentle slope. Steep leeward flanks are determined by the stable angle of repose for the sands, and by the eddying influence of winds curling around the dunes. Successive beach and dune ridges may be seen to differ from one another in width and crest line delineation. This is to be expected from the formation processes described. New accumulations of shore debris, such as a storm may deposit, might promote the formation of a new ridge before an earlier landward one has acquired significant growth. Thus, the new dune or ridge may achieve considerable stature before a newer dune in turn checks its growth. Additionally, occasional shoreline retrogradation has been observed to combine several low dunes into one of greater size and height.

The alternate elevation and size of a beach dune or ridge is due to great variability of marine forces. Several of the important factors so influencing dune ridges are summarized in Table 4-1.

Table 4-1. Some Factors Operative in Dune Altitude

Factors favoring increase of altitude (Building up, construction)	Factors opposing increase of altitude (Tearing down, destruction)
Wind	
Moderate or weak air currents transport sand upward the windward slope. Aggradation	Stronger upper air currents remove sand from top. Erosion
Gravity	
	Favors erosion and is in direct relation to density of particles moved. Opposed aggradation.
Sand Supply	
Continued sand supply In turn depends on (a) Continued shore current replenishing sands cast ashore. (b) Erosion elsewhere to furnish sand to current. (c) Strong wave action to wash up and deposit sand on higher back beach.	Terminated sand supply.

Sand Supply (continued)

In turn depends on (d) Deep offshore bottom favoring strong surf or an exposure to open sea. (c) Age of shore, young shores having little or limited supply, while older ones tend to have a more copious amount.	
Time	
Prolongation of time of action obviously tends to increase the height—slight settling disregarded.	Converse evident
Exposure	
Open exposure favors strong wave and wind action.	Shelter reduces power of waves and wind and therefore their ability to transport sand.
Beach Stability	
Stability of beach, actual or relative results in long continued actions which concentrate sand on a given dune ridge.	Widening of beach or prograding shoreline results in new foredunes which in turn (a) Reduce the time of action on any one ridge. (b) Intercept sand headed for the back ridge.

Reprinted by permission of the Bureau of Geology, State of Florida Department of Natural Resources

It is clear that for continued building and longevity of large dune ridges, abundant plant life is required to anchor the shifting sands. Dunes anchored solely by annuals like beach orach, sea rocket, or saltwort seldom last a year. The pioneer dune builders are the perennial herbs, represented by sea oats, marsh elder, and sea purslane. Many herbs that are themselves incapable of building or holding a large dune, live and thrive in sand hills captured by the pioneering vegetation. Normal dune plant succession replaces the pioneer plants with intermediate vegetation such as saw palmetto, sea grape, broom sedge, and beard grass. Ultimately these plant associations are replaced by such shrubs or trees as Rolf's oak, buckthorn, and shore bay. These latter types are usually found growing on the most landward ridges, as shown in Figure 4-5.

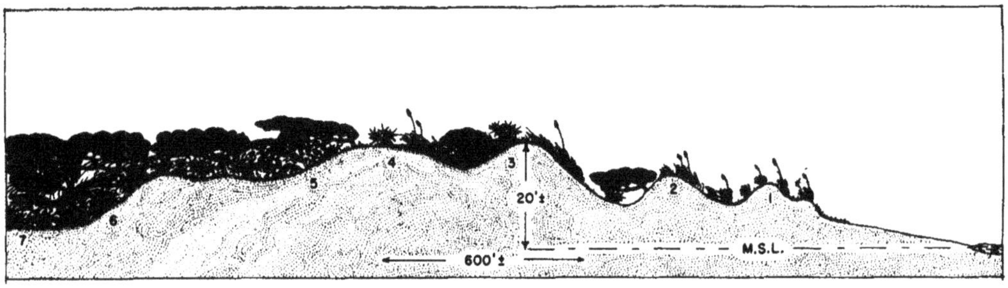

Figure 4-5. This profile of Crescent Beach is typical of the vegetation progression of coastal dunes. Youthful foredunes (1) and (2) are protected by sea oats. The middle dunes (3) and (4) reflect an intermingling of saw palmettos and low shrubs with well-established sea oats. Progressive plant succession on older inland dunes (5) and (6) include oak trees whose shade will eventually destroy many of the low shrubs and palmettos. (Reprinted by permission of Florida Bureau of Geology, State of Florida Department of Natural Resources.)

Beach dunes are retrograding in some areas due to "dune erosion." Such erosion normally occurs on shores which have remained stationary for extended periods, have emerged very slowly, or those that are actually submerging. In its beginning form, dune erosion appears as a small scarp marking a slight encroachment of the seas. Progressive erosion by undermining waves brings disaster to cabbage palm and oak-lined ridges. Ponte Vedra Beach (north of St. Augustine) is an exemplary modern area of dune erosion.

Ponte Vedra also presents a spectacular view of a second destructive dune process, "blowout." Blowouts occur where strong winds find a breach in the dune's defensive vegetation. The breach may be as inconspicuous as the death of a small plant whose roots have grasped only small amounts of sand. Strong winds whip away the unprotected sands and widen the breach. Soon adjacent plants and shrubs are undermined and topple, exposing progressively larger areas to wind erosion. The blowing sands are deposited upon the more landward dunes and ridges at a rate which smothers their protective vegetation. This destruction and overwhelming of dunes continues until they have been leveled or until the vegetation can reassert itself during seasonal wind fluctuations (see Figure 4-6A-D).

Many of Florida's finer dunal areas have been destroyed by bulldozers and construction. This practice initially received attention in the tourist areas surrounding the Miami area but in recent years has expanded to all of the state's coastal areas. The state government has purchased several areas and converted them into parks and natural preservation areas, so that these dunal areas may be understood and enjoyed by all her citizens.

Man as a Geomorphic Agent

From our previous discussion it has become apparent that man is also an agent of change to the surface of Florida. Man's changes to the face of Florida have evolved gradually but with progressively increasing intensity. We can attribute these factors to: (1) an ever increasing population, (2) an increase in man's control and use of energy and technology, and (3) the demands and conditions of urbanization. The effect of man's influence extends into every corner of Florida, even to the depths of Big Cypress Swamp and the wetlands of the Everglades. Human influences are seen in such areas as deforestation, manipulation, and movement of earth and soil; changes in the water regimes of streams, lakes, and seashores; and even the geology of the state.

Each area of Florida possesses what can be called a geologic norm of soil erosion which is controlled by the natural condition of climate, soils, and prevailing bedrock. However, where man has deliberately cut the forests or removed the grass and palmetto cover, erosion becomes accelerated. Wolman's (1957) studies of erosion indicated that construction of subdivision housing and roadways increased the tonnage of sediment removal per acre 20,000 to 40,000 times the geologic norm of natural woodlands.

The major portion of sediments lost each year comes from Florida's farmlands. This is attributed to the fact the Florida's cultivated and grazing acreage far exceeds the acreage devoted to urban and mining activities combined. Over seven million acres in the state were devoted to farm activities in 1978.

Man is also carving the Florida landscape into new landforms through his surface mining activities. A 1967 survey on surface mining activities conducted by the U.S. Department of the Interior estimated Florida had 180,000 acres devoted to strip mining, spoil banks, and mine access roads. Florida ranks fourth in the nation in total acreage disturbed by surface mining. Most of this land is devoted to the mining of phosphates. In the intervening years since the USDI survey, approximately 30,000 acres have been reclaimed by Florida's phosphate companies. Most of the reclamation has occurred near urban areas where the improved lands have been sold, probably for housing and commercial development.

Strip mining in Florida's flat and rolling terrain can be compared to the plowing of farm furrows, with the overburden overturned to form ridges parallel to the cut. The stripping techniques create a ridge and valley landform which resembles a giant washboard. Ridge crests vary from 2 to 50 feet or more above the valleys, with slopes of 17° to 39°. Slope angles are determined solely by the repose angle of the overburden. The valley widths range from 50 to 100 feet, depending upon the working characteristics of the earthmoving equipment. The surface materials of the ridges consist of fresh strata which formerly lay directly above the deposit. The fresh unweathered materials tend to produce initially steep, easily eroded ridges as shown in Figure 4-7. Planned tandem use of shovels and draglines can result in placing the original surficial materials atop the spoil ridge, thus simplifying reclamation and promoting revegetation.

Figure 4-6A. Embryonic sea oat dunelets and convex dune covered with railroad vine along Cape Canaveral Beach.

Figure 4-6B. Rear dune area with growth of oak, saw palmetto and sea grape.

Figure 4.6C. Oak scrub, cabbage palm, saw palmetto, and railroad vine protect 25-foot high dune on Manhattan Beach, Duval County.

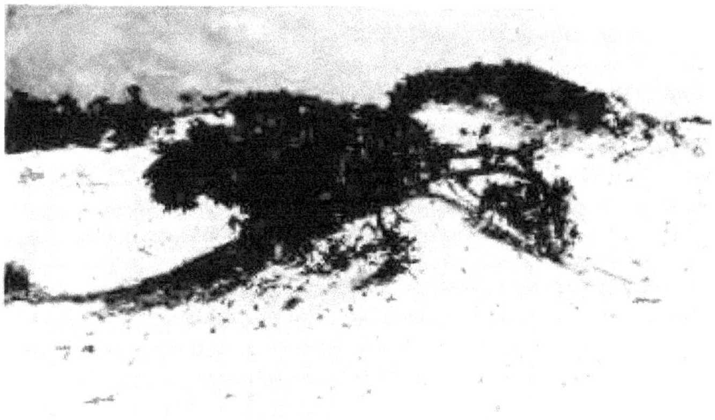

Figure 4.6D. Magnolia dune remnants in blowout along Panama City Beach (Bay County). (Reprinted by permission of the Bureau of Geology, State of Florida Department of Natural Resources.)

Figure 4-7. Phosphate strip mining at the Agrico mine in Hillsborough County.

PHYSIOGRAPHIC DIVISIONS OF FLORIDAN PLATEAU

We have studied the factors that chiseled the form and topography of Florida. We can now divide the state into physiographic or landform divisions based upon relief and surface features differences. While fineness of detail may be used to divide the state into 50 or more such regions, simplicity directs the use of only four generalized physiographic divisions. These include (1) the Northern Highlands, (2) Central Highlands, (3) Marianna Lowlands, and (4) Coastal Lowlands. These divisional areas are reflected in Figure 4-8, while the major topographic features are shown in Figure 4-9.

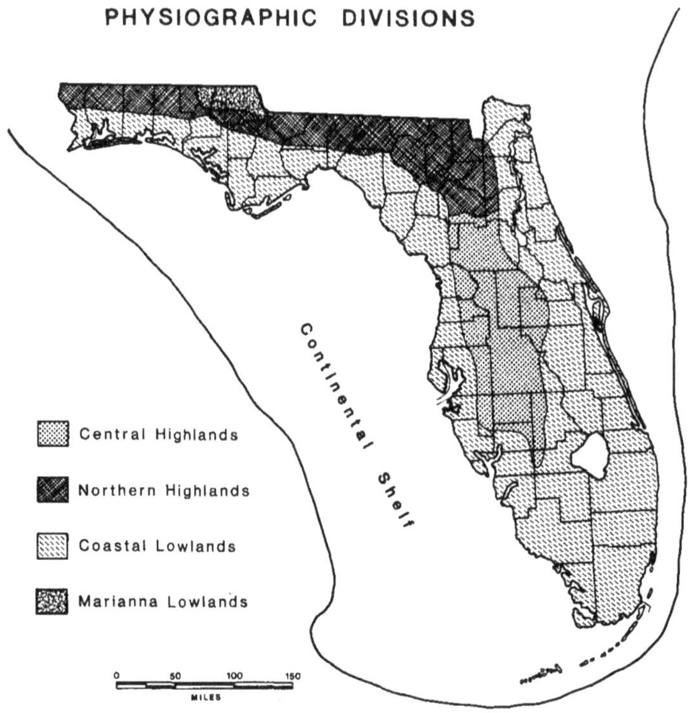

Figure 4-8. Physiographic divisions of Florida.

The Northern Highlands

The Northern Highlands represent an almost continuous hill and ridge area crossing the northern portion of Florida. On the east the highlands begin at Trail Ridge (Baker-Putnam Counties) and extend north and west into Georgia and Alabama.

The Northern Highlands are separate from the numerous ridges of the Central Highlands only because the latter region is more dissected than are the relatively intact Northern Highlands. Both Northern and Central Highlands apparently once comprised a single, integrated highland which has since been partitioned by erosion and groundwater solution activity. Actually the division between the Northern and Central Highlands is not prominently marked by any important surface feature, but rather by a series of relatively unimportant gaps and valleys. These dividing features include High Springs Gap, Alachua Lake Cross Valley, and Kenwood Gap—all of which are only 8 to 12 miles wide.

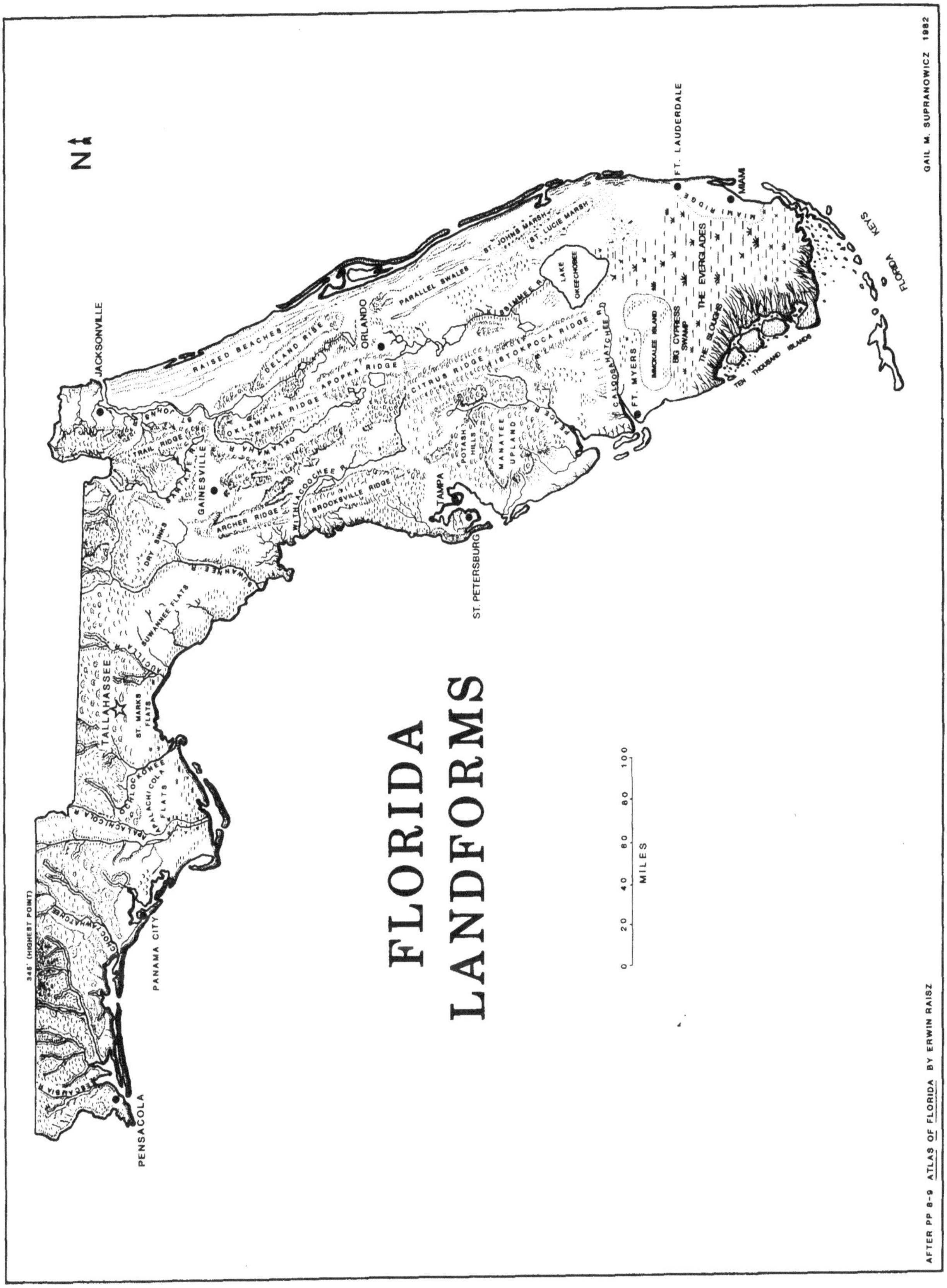

Figure 4-9. Major topographic features of Florida.

However, the Northern Highlands are significantly bounded along the eastern and southern margins by an outward facing scarp which extends regionally from the Atlantic coastal plains through the breadth of the Panhandle and into Alabama. This scarp is known as the Cody Scarp in Florida, and is the most continuous and persistent topographic feature traversing the state. It extends some several hundred miles, and is unbroken except by major streams. In many places it drops sharply 100 feet in less than one-quarter mile; in others it is marked by only a general flattening of terrain. On the east the Cody Scarp comprises the seaward or eastern side of Trail Ridge, where it extends some 125 miles from Putnam County to the St. Marys River. The eastern scarp of Trail Ridge has been shaped primarily by marine erosion. The scarp swings westward near Palatka following a terrace line of possible Wicomico Seas. This transpeninsular portion of the scarp appears to have been shaped primarily by stream erosion and solution. As the scarp proceeds westward into the Panhandle, marine erosion again seems to have shaped the features of the scarp.

Many of the streams flowing west of the Tallahassee Hills emerge from unusually steep heads. This feature has given rise to such descriptive names as Mossy Head, Deep Head, Bear Head, Deer Head, and White Head. The steepness of these stream heads is attributed to a combination of their extreme youth (most appeared years after the ice ages) and the underlying porous sands of the region. The sands absorb most of the rainfall, resulting in little surface erosion. Headward stream growth is attributed to the springs which gush from the hillsides and gradually eat away into uplands.

A large portion of the western area (termed Western Highlands) lies in the pine woods covering the steep hills of Eglin Air Force Base, formerly Choctawhatchee National Forest. Lumber and naval stores are the principal natural products of this region. Farming is more prevalent along the rolling and more gentle slopes of the Tallahassee Hills.

East of the Tallahassee Hills lies a nearly level upland plain deposited as a portion of the Sunderland Terrace. The plain proceeds northward into Georgia where it includes the Okefenokee Swamp. The eastern end rises along Trail Ridge, which stood as a sandy peninsula and bar in interglacial Sunderland Seas.

A notable landmark of the Northern Highlands vanished in the last century. Throughout recorded Florida history a mysterious plume of smoke rose continuously from the watery jungle of the Wacissa Swamp. Local Seminole legends said the smoke had always been there. Eighteenth and Nineteenth century navigators used the plume as a beacon in sailing the Gulf coastal waters. Local settlers referred to the smoke cloud as the "Wacissa Volcano" although none are known to have entered the swamp to study it. The swamp, it seems, was home to a band of Seminoles who intermittently raided surrounding settlements. No settler was interested enough to challenge their rule to see the "volcano" closely. In 1886 the Tallahassee Hills were rocked by an strong earthquake felt throughout the Southeast. The quake and its tremors apparently sealed the source of the Wacissa plume, for its black and grey spiral vanished with the event. Numerous attempts to locate a volcano site in this area have proved fruitless.

Marianna Lowlands

The Marianna Lowlands occupy a relatively small rectangular area (approximately 30 by 64 miles) in Holmes, Washington, and Jackson Counties. The lowlands extend north into Georgia and Alabama along valleys of the principal streams (particularly the Chattahoochee River*). The lowlands are bordered in Florida entirely by the Northern Highlands. This border is marked along the entire distance by the steep north-facing Holmes Valley Scarp.

In the geologic past these lowlands constituted a portion of the continuous highlands traversing northern Florida. The area contains numerous abandoned river valleys and abundant limestone cavern systems. These features, in conjunction with the distribution of stream terraces and the wide gap along the Choctawhatchee River, suggest that the lowlands

* This is in Florida.

are the result of erosion by the Apalachicola, Chattahoochee, Chipola, and Choctawhatchee Rivers.

The Marianna Lowlands is a region of low rolling hills and hollows. It is dotted with sinks, which range from steep-walled holes like Falling Water (near Chipley) to broad, shallow depressions. Many of the sinks contain ponds or small lakes which are thickly ringed with cypress trees and other waterloving vegetation. Other sinks differ in appearance from the surrounding higher land only by their darker, more moist soils. Most of these sinks occur where the Ocala and Marianna limestone lies near the surface or are actually exposed. A well-developed dendritic stream pattern drains the lowlands.

The main transportation artery of the area follows the Old Spanish Trail and contributes to the prosperity of the region's agriculture.

The Central Highlands

The Central Highlands represent the remains of a former elevated and continuous terrace. Today the remnants remain only as ridges preserved by some resistant attribute obtained during their deposition (i.e., localized properties of the Hawthorn formation) or predissection existence. Major highland areas include the Brooksville, Winter Haven, and Orlando Ridges. These and other lower ridges enclose such broad lowland areas as the central and western valleys and the St. Johns River offset. The general pattern of both ridges and valleys trend linearly parallel with the Atlantic coast. Many of these features are long and narrow. The Lake Wales Ridge, for example, is only a few miles wide and over 100 miles long. The highest point is said to be Iron Mountain (325 feet above sea level) in Polk County, on whose summit stands the carillon called the "Singing Tower."

Ridges, for the most part, are surfaced with well-drained sands and are appropriately called **sand hills.** Groundwater action and karst topography is common in the lower elevations, where fractured overburden facilitates the formation of springs and sinkholes. The Central Highlands have also been termed the Lime-Sink District for this reason.

Central Florida is the center of Florida's citrus growing industry. The hillsides above the lakes are checkerboarded with evenly spaced rows of orange and grapefruit trees whose glossy leaves and golden fruit are reflected in the lake waters. Along the western edge of the highlands the hills become lower and more interspersed with broad low plains and a lacy pattern of innumerable shallow lakes, including Lakes Tsala Apopka, and Panasoffkee. The abundance and intricacy of the lake patterns is attributed to the extremely shallow Ocala limestone. Many of the lakes are surrounded by extensive flats or "prairies."

The Coastal Lowlands

The Coastal Lowlands are generally low in elevation and poorly drained. The topography is characterized predominantly by marine activities and terrace developments. Thus, most of the surface features generally parallel the coast. Those lowlands lying east of the Central Highlands are called the Atlantic Lowlands, while those to the west of the highlands are referred to as the Gulf Coastal Lowlands. To the south of the highlands lie the distal lowlands which have a genetic relation to both the Atlantic and Gulf. Both the Gulf and Atlantic Coastal Lowlands are relatively narrow and average 30 miles wide. The lowlands incorporate four terrace levels, and in most cases are bounded on the landward sides by the Wicomico terrace at 100 feet mean sea level.

The topography and drainage of the Atlantic Coastal Lowlands is influenced primarily by select barrier islands, lagoons, estuaries, relict spits, and bars, with intervening coast-parallel valleys. The St. Johns River traverses 250 miles along the path of an ancient lagoon. The river's course parallels the coast nearly the entire distance before turning sharply into a relicit estuary to enter the Atlantic. The Banana and the Indiana Rivers also flow to the coast. Actually, both are lagoons which receive some fresh water from small streams that drain into them. The landforms of the Gulf Coastal Lowlands reflect the development and influence of sand dune ridges, coastal ridges, and karst topography. The lowlands persist as gently sloping but undulating plains. The gentle undulations and hills are the remains of dune and coastal ridges which wander parallel to the coast. Sinks and springs are abundant in the Gulf Lowlands due to the near-surface Ocala limestone which underlies the area.

Both Atlantic and Gulf Coastal Lowlands are covered by pine forests with an undergrowth of saw and cabbage palmettos. Cypress swamps are interspersed throughout the lower, poorly drained areas.

The distal lowlands encompass the southern end of the Florida peninsula and include at least one-third of the state. This area of the state contains a variety of outstanding topographic features, among which are: (1) the low relief of the Everglades-Lake Okeechobee Basin and the many changes that have occurred in the levels of the surface of the organic soil in the Everglades since the draining of the parts of this region; (2) the low relief and inundated condition of most of the southern coast and the southwest coast regions; (3) the relatively high relief of the highlands ridge and parts of the Atlantic coast strip and Miami ridge, the latter two acting as a barrier between the Everglades Basin and the Atlantic Ocean; (4) the slightly undulating and partly dissected plains of the flatlands regions that are the remains of old sea terraces eroded to their present condition by a few streams and rivers; and (5) the confused topography of the Big Cypress region and some of the southern Everglades where local weathering has produced numerous irregularities in the land surface (Davis, 1943).

The Everglades and Lake Okeechobee represent a somewhat unique former seabottom. The region became a basin when anticlinal folding moderately elevated the lands to the south. In the regime of a wet climate, Lake Okeechobee filled with water and became a lake. Because the lake is shallow and elevations to the south are only a few feet higher, overflow waters from the lake drained to the south. The resulting environment was a sea of grass—called Payhokee (Figure 4-10) by its Seminole inhabitants. Drainage of the area by Miami, West Palm Beach, and other canal systems have resulted in drying of the swamp's muck soils and soil subsidence affecting some 250,000 acres. The man-induced subsidence has reversed the outward, natural drainage around Lake Okeechobee, and the lands now slope toward it.

Figure 4-10. Payhokee or "sea of grass" in the Everglades.

Drainage of the Everglades area for agricultural purposes began in earnest in 1906. Manmade drainage networks for the Everglades are shown in Figure 4-11. The rich muck land is extremely fertile and crop productive. However, dry organic murk is very susceptible to fire, which burns down into the soil. Muck fires can normally be quenched by heavy rains, from lack of oxygen, or encounter with the water table.

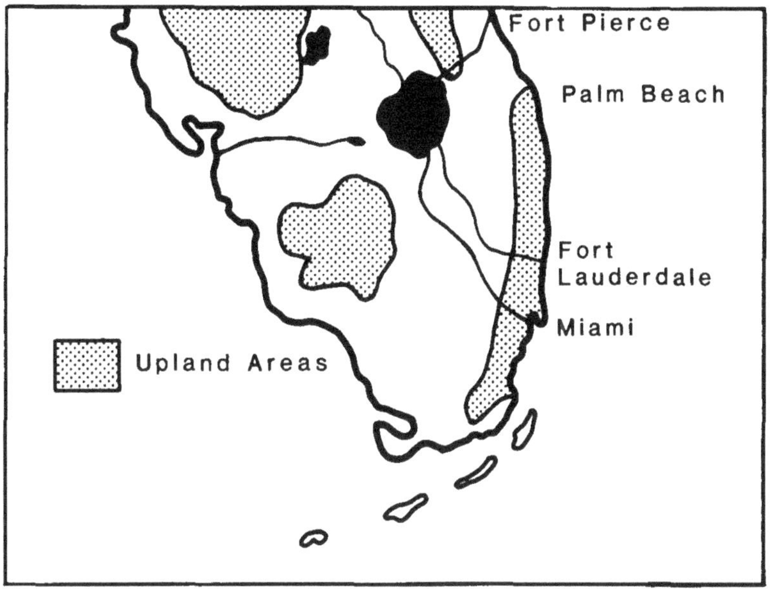

Figure 4-11. Drainage system of the Okeechobee-Everglades Basin. The lake is fed by the Kissimmee River to the north. Manmade drainage canals also carry water to Atlantic Coastal cities.

In contrast to its flatland topography, the distal area also contains a series of elevated rocklands, sand ridges and dunes, lagoonal valleys and swales along the Atlantic coast strip and Miami rock ridge. These features were developed by advancing seas during a Pleistocene interglacial period. The combination of dune ridges and Miami oolitic limestones produced a highland barrier between the Everglades Basin and the Atlantic Ocean. Waters from the Everglades and ocean alternately cut across these uplands to form transverse valleys and a very uneven topography. Those transverse valleys which contain rivers and streams are known locally as transverse glades.

An elevated area rising some 40 feet above the lower flatlands is found in DeSoto County. This area, called the Immokalee Rise, formed as an island in the Talbot Seas. One can still see the ridges and pond valleys which are the remains of the bars and swales formed along the shores of that ancient sea.

The Big Cypress region is a flat, swampy area festooned with orchid-studded, moss-hung trees. Local elevation varies from 5 to 20 feet mean sea level. The general, nearly flat topography results in poor surface drainage, which is further confused by irregularities in the soils and surface rock. Drainage along the western portion of the region is through cypress sloughs toward the south, while drainage along the eastern areas is toward the Everglades. The Big Cypress also is dotted with solution holes and there is evidence of underground drainage. Many of its shallow depressions are seasonal ponds. The main drainage appears to occur through elongated swamps and slough marshes.

The Florida Keys comprise a long fringe of low islands that curve south-westward along the edge of the Straits of Florida from Biscayne Bay to Key West. Outlying islands such as the Marquesas Keys, Rebecca Shoal, and the Dry Tortugas might be included with them. The outer fringe of keys, terminating at

Loggerhead Key, is an emerged coral reef (Key Largo limestone) composed of limestone containing coral heads in their original position. Most of the keys are long and narrow and are strung out in a line. Other keys, known as the Pine Islands, are more irregular in shape and are composed of oolitic limestone (Miami oolite) like that of the mainland. The Marquesas Keys and the Dry Tortugas consist of "coral sand" heaped up by the waves in the form of atolls on a slightly submerged bank. All of the keys are very low. Probably none exceed 15 feet above sea level.

Mangrove swamps occupy much of the shallow water behind the keys. The head of Florida Bay is dotted with them and they fringe the mainland from Biscayne Bay as far west as Cape Romano. Most of the Ten Thousand Islands consist of mangrove swamps, though some of them are partly submerged sand dunes. Mangroves grow in sheltered tidal waters as far north as New Smyrna Beach on the east and Cedar Key on the west. Their northern limits is determined by climate, for they cannot endure much freezing weather (Cook, 1945).

DRAINAGE OF THE FLORIDA PENINSULA

An unusual facet of the peninsula's surface drainage system is that the state's major divide does not follow the highlands running the length of the peninsula. Rather, it runs across the state at its widest point (essentially from Cape Canaveral on the east to St. Petersburg Beach on the west). This divide separates the northward flowing streams and rivers from those flowing southward. The major north-flowing rivers include the St. Johns and its tributary; the Oklawaha; the Econlockhatchee; and to the west, the Withlacoochee. The main south-flowing systems include the Peace, Kissimmee, and Caloosahatchee Rivers.

The divide is believed to have resulted from side-by-side placement of rocks of differing solution resistance, by erosion of fault zones crossing the peninsula, or by a combination of these causes. Evidence for these beliefs lies with two well established facts: (1) rocks north of the divide are primarily soluble limestones brought to the surface by the Ocala uplift, south of the divide the rocks consist essentially of insoluble elastic sandstones and conglomerates; and (2) a series of parallel northeast-southwest trending faults crossing this section of the state (though no one fault is recognized as completely crossing this section of the state in this area).

It is interesting to note that the north-flowing streams are not apparently diverted from their courses by entrance into subterranean routes (with steeper gradients) flowing directly to the sea. These streams have apparently resisted underground capture because their valley floors are below the piezometric level and in many cases are actually areas of artesian discharge. The latter concept is evidenced by the presence of large springs that provide additional waters to the transversing streams. The major peninsular streams all flow parallel to the coast. The stream channels in most cases once existed as long narrow lagoons, or bays, behind offshore bars and islands. According to some investigators, these lagoons probably resembled the present Indian River along the east coast today. These lagoons were probably inundated to varying depths and partially filled with sediments during interglacial periods: during glacial times the areas drained and rivers reestablished their courses. Additionally, it is common today to find parallel lines of swampy sinks, with varying amounts of drainage, appearing between many old beach ridges.

The only significant exceptions to the general parallel coast drainage is found in the northwest. Here, several short streams flow essentially perpendicular to the coast. Only two extended major streams are consequent (follow the direction of the initial slope on emerging in Florida) to the area. These are the Suwannee and Chattahoochee Rivers, which rise in Georgia. Because the peninsula was significantly shortened and narrowed during the high sea level periods, all other extended streams from Georgia were prevented from penetrating the peninsula. Florida receives approximately 55 inches of rainfall annually. This rainfall may be translated into nearly 150 billion gallons per day. Of this rainfall, some 40.5 billion gallons reach the ocean through stream discharge. The remainder is lost to evapotranspiration (100 billion gallons) and the underground flow (10 billion gallons), respectively. Additionally, some 25 billion gallons of water enter the state daily, delivered by streams originating in Alabama and Georgia. Discharge by the state's major streams is reflected in Figure 4-12, while one of many lesser streams, the Santa Fe, is shown in Figure 4-13.

Florida's flat terrain does not promote the rapid flow rate observed in streams with steeper gradients. The rivers are relatively shallow, and for the most part banks are not high. The Oklawaha, for example, flows through Lakes Tsala Apopka, Dora, Eustis, and Griffin before changing direction and entering the St. Johns River. During periods of flood, overflow water from a variety of rivers fills lakes occupying abandoned river meanders. Such lakes are termed **meander lakes.** Meander lakes of the St. Johns system include Lakes Washington, Harney, and George.

Florida's lakes and wetlands are important contributors to the state's hydrologic system. They act as water storage basins which regulate the flow and discharge of streams. Each basin is capable of retaining and storing water either because it is underlain by impervious clays and organics or because the floor of the basin lies below the water table.

Also of notable interest is the termination of five important terrace ridges at this divide. The Brooksville and Orlando ridges run north from this point; the Lake Wales, Lakeland, and Winter Haven ridges extend south. While all five ridges are essentially parallel, none is an extension of another. These ridges in turn become divides separating individual basins of several important streams whose headwaters abut the cross-state divide.

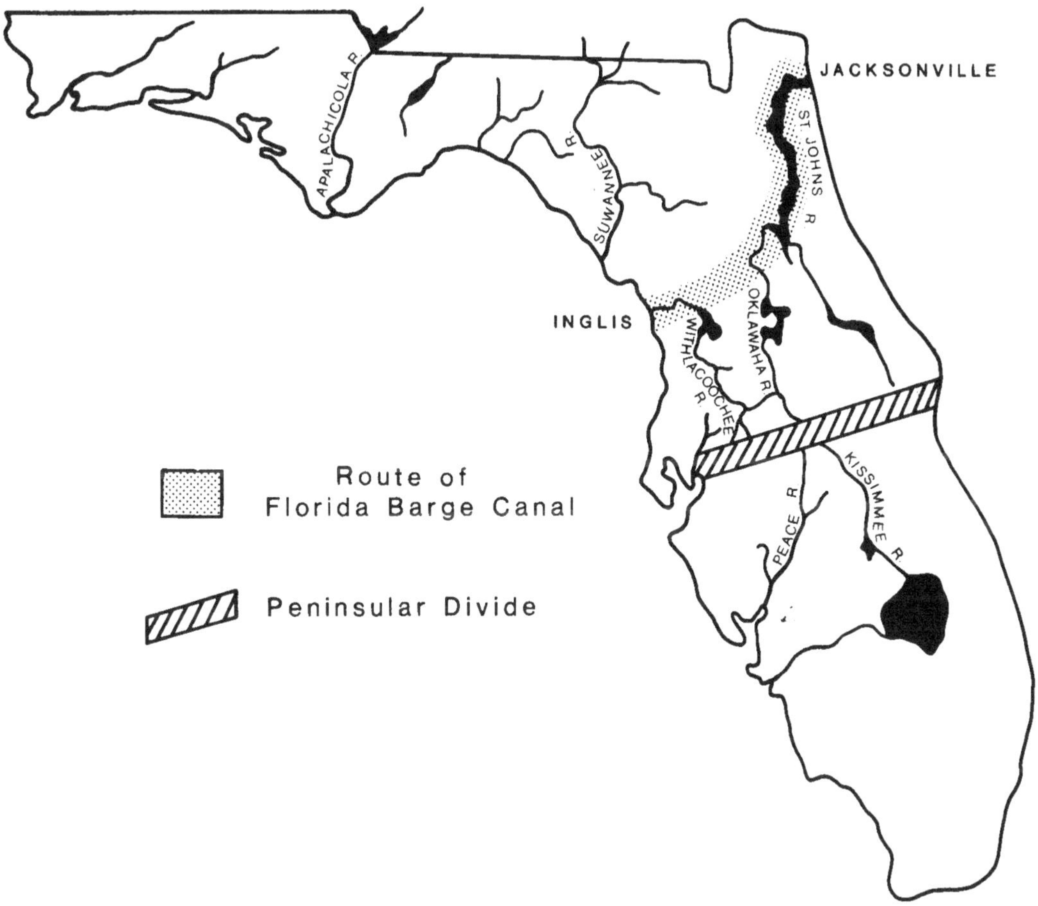

Figure 4-12. Florida's major drainage systems. The Peninsula Divide separates the north and southward flowing steams.

Several north-flowing streams follow troughs in the bedrock related to the Ocala uplift. The Oklawaha, however, rises in a zone of relict beach ridges and follows their patterns through an area extensively altered by solution action. Consequently, the Oklawaha wanders through a long chain of lakes before turning east into a fault where it joins the St. Johns.

The pathway of each river is unique (Figure 4-14) as it wanders among tree-lined banks and across the landscape. However, many common features may be recognized and used as a basis for dividing the streams into descriptive categories. At least three such categories are evident: alluvial, blackwater, and spring-fed.

Alluvial rivers transport large quantities of eroded sediments. The majority of Florida's alluvial rivers originate in Georgia and Alabama and transport clays and silts from these areas, (i.e., the Apalachicola and Choctawhatchee). Because of their sediment loads, alluvial rivers usually appear muddy and brown. These rivers routinely flood over their banks and deposit their sediments along their banks and floodplains. Natural levees are common along their banks where sediments have raised the ground level along the waterway. Low sloughs and ponded depressions lie landward of these levees. Hardwoods such as oak, hickory, and maple are common along the levees. The flood-prone bottom lands (often with standing water) are often called backwaters and are dominated by such water-tolerant trees as cypress and ash.

Blackwater rivers receive their name from the staining effect of oak tree tannin and organic acids leached from pine forests. Blackwater rivers are the most common type found throughout the state and originate primarily among the hills of the central highlands. The Econlockhatchee, Blackwater, and Peace Rivers well represent this classification. Because these rivers commonly originate among forested hills, sediment levels are low and river bottoms are usually lined with muck/sand/clay mixtures. Trees along the waterways are usually the more water-tolerant types found in backwater swamps, although oak and bay trees often are found on higher ground. Understory shrubs are abundant where flooding is seldom significant.

Spring-fed streams are generally clear and flow through relatively narrow floodplains. These streams maintain an even flow rate year round and seldom experience flood conditions. Because of this, erosion sediment levels are insignificant although dissolved carbonate levels tend to be moderate (200-400 ppm). Vegetation resembles that found along blackwater streams.

Many of Florida's larger rivers express a combination of the characteristics described. The various features appear along individual runs or segments of the waterway as springs provide crystal waters to an otherwise blackwater current.

Quicksand pits are occasionally found along the riverbanks and around small springs. **Quicksand** is a sand-water mixture in which the sand grains temporarily float. Until the sand grains settle the suspension flows as a thick liquid. Semi-permanent pools of quicksand can form where sand overlies an upwelling spring. The upward pressure of the water prevents the sand from settling. Short lived pools often form along streams where loosely packed sands are saturated with water and then subjected to a strong vibration, such as an earthquake or man's mechanical equipment produces. However, such quicksand dangers are overrated. Drowning is difficult in quicksand as the human body is more buoyant in quicksand than in a swimming pool.

Many of the peninsula's wetland bogs and swamps are the evolutionary offspring of former lakes and lagoons. The life span of lakes and lagoons is relatively short by geological time standards. Their wetland evolution is brought about by (1) basin filling due to accumulation of sediments carried by inflowing streams or slumping and erosion of basin (often sinkhole) sides; (2) formation of basin drainage channels by overflow, erosion, or manmade methods; and (3) accumulation of organic mucks produced by anaerobic decomposition of vegetation. Many of the lakes lack surface drainage channels and thus evolve into bogs and swamps as their subsurface drainage is reduced.

Figure 4-13. A view of the Santa Fe River at Olena State Park. This stream is a tributary of the Suwannee River.

Figure 4-14. The Withlacoochee River as it meanders through the Florida landscape. (Reprinted by permission of Florida Game and Fresh Water Fish Commission.)

Man's Impact on Surface Drainage

The main effects of urbanization are seen in the reduction of rainwater infiltration and the increase in surface runoff. The construction of buildings, pavements, and parking lots is resulting in an ever-increasing area of impervious land surfaces. Studies of residential land surfaces estimate that construction on small lots (6000 sq. ft.) reduces infiltration about 80 percent, while on larger residential properties (15,000 sq. ft.) only some 25 percent of the surface is rendered impervious. Reduction in infiltration produces a corresponding reduction in ground water recharge and thus groundwater's contribution to stream flow. Base flow (or normal dry season stream level) is lowered correspondingly.

Reduction of surface permeability inevitably results in increased overland flow of rainwater from urban areas. The impact of urban flooding is not restricted to roadways and lowland areas but is also becoming a major influence on flow in natural stream channels. The use of sewer systems introduces storm water runoff directly into the lakes and rivers without providing an opportunity for infiltration. The combination of increased overland flow and storm sewers conspire to produce an increase in both flood frequency and magnitude. The flooding is now acute in many suburban communities where certain low-lying residential areas formerly free of flooding are now regularly inundated by overflowing streams and lakes.

Many Florida communities are developing solutions to the problems described. In the Orlando area storm water runoff in new residential areas is diverted into holding basins where it can eventually infiltrate into the sands or evaporate. Drainage wells are also provided around many of the Orlando area lakes to prevent overflow. The drainage wells allow excess waters to drain directly into the underlying cavernous limestones.

Most Florida communities exercise floodplain zoning laws to restrict development of hazardous areas. The resulting green belts in the flood plains are becoming parks and wildlife sanctuaries.

Many areas of Florida are subject to flooding on a massive scale. During March of 1960, 10,000 square miles between Orlando and Tampa were severely flooded. The flood resulted after a not uncommon heavy rainfall (10 inches in three days) Occurred when most area streams were already near flood stage. Damage was excessive.

To reduce subsequent massive flooding in central Florida the Corps of Engineers has developed the Four-River Basin Project. The plan is to provide water control and flood protection for a 6,000 square mile basin that includes the headwaters of the Withlacoochee, Oklawaha, Peace, and Hillsborough Rivers. Average flood frequency in the basin occurs once every 10 years. The plan includes nine major artificial storage reservoirs, increased storage in the area's major lakes, and enlarged stream drainage capacities.

In 1968, the Corps of Engineers also began construction of the Cross Florida Barge Canal. The Corps cleared some 5,500 acres of hammock along the Oklawaha River by pushing over the trees and crushing them into the mud. Few trees were salvaged for lumber. Closing of the Rodman Dam created a reservoir which flooded 13,000 acres. Because reservoir water levels were stabilized for navigation and flood control purposes, the reservoir lacked the natural seasonal fluctuation of water levels necessary to maintain high water quality. The reservoir waters are nutrient rich, having been located in a fertile river swamp forest. The reservoir is now showing signs of rapid aging or eutrophication including (1) excessive aquatic plant growth; (2) algae replacement of higher order plants; (3) bottom buildup of organic muds and muck; (4) low levels of dissolved oxygen, with resultant decline in fish population; and (5) high turbidity levels.

Work on the Cross Florida Barge Canal ended in 1972. Fortunately, only about 10 miles of the Oklawaha System have been adversely affected.

Chemical pollution by direct disposal of industrial wastes from Florida's factories is minimal. In the 1980's the Iron Bridge Sewage Treatment Plan east of Orlando dumped nearly a million gallons of treated sewage in the Econlockhatchee River daily, due primarily to inefficient plant operations.

Agricultural chemicals (both fertilizers and pesticides) are the major pollutants of the Everglades. These chemicals enter the water stream from large sugar farms lying north of the swamp lands. Efforts are now being made to correct these problems, and sugar growers claim their efforts have improved the water quality to be better than that of natural rainfall. Problems in the Everglades also are related to massive canals and drainage systems originally intended for flood control.

Phosphate mining releases a variety of chemical wastes, including some low-level radioactive contaminants into the hydrologic system.

The: **Preface of Restoration Ecology** (special issue: Kissimmee River Restoration) provides the history and restoration of this major Florida river:

> The historic Kissimmee River meandered blindly for approximately 166 km within a 1.5-3-km-wide floodplain. Between 1962 and 1971, a 90-km long, 9-m deep, box-cut canal was dredged through the river/floodplain system. The canal was divided into five reservoir-like impoundments, called pools, by a series of six water control structures (i.e., gated spillways), which regulate water levels and flow in each pool.
>
> The excavation of the canal and placement of spoil obliterated ⅓ of the river channel and 2800 hectares of floodplain habitat. These and other physical modifications, including altered inflow regimes and drainage of ⅔ of the historical floodplain.
>
> The Kissimmee River restoration project seeks to undo these impacts by eliminating two of the water control structures, removing the spoil that currently covers portions of the floodplain and using it to fill 335 linear kilometres of canal (re-exposing buried floodplain soils), and re-carving pilot channels to reconnect disjunct sections of the remnant river channel. These measures will restore the historic form and hydrology and associated ecological resources of 70 kilometres of continuous river channel and 11,000 hectares of floodplain within the central portion of the channelized system.

REFERENCES

Arrington, D. Albrey (1995), *Restoration Ecology*, Journal of the Society of Ecological Restoration, Vol. 3, No. 3, 9/95.

Brown, M.W. "New Clues to Earth's Largest Extinction," *New York Times, Orlando Sentinel*, January 24, 1993, p. G-7

Carpenter, R. H. (1970), "Metamorphic History of the Blue Ridge Province of Tennessee & North Carolina," *Geological Society of America* Bull. Vol. 81, pp. 749-762.

Cooke, C.W. (1945), *Geology of Florida*: Florida Geol. Survey Bull. 29.

Davis, J. H. (1943) *Natural Features of Southern Florida,* Florida Geol. Bull. #25.

Dietz, R.S. & J.C. Holden (1970), "reconstruction of Pangaea: Breakup and Dispersion of Continents, Permaian to Present," *Journal of Geophysics Research*, Vol. 75, pp. 4939-4956.

D'Oliverira, S., "Here It Comes," *Orlando Sentinel*, October 20, 1996, p. 7, Discover Channel, "3 Minutes to Impact," program February 19, 1997.

Emmons, W.H., I.S. Allison, C.R. Stauffer, C.A. Thiel (1960), Geology: Principles and Processes, McGraw-Hill Book Co., New York, p. 491.

Flazel, C.T. (1986), "Asteroid Impacts, Sea Floor Sediments, and Extinction of the Dinosaurs," *Sea Frontiers*, Vol. 32, pp. 169-178.

Forest H.E. (1935), *The Atlantean Continent, Its Bearing Upon the Great Ice Age and the Distribution of Species,* 2nd ed., London.

Hatcher, R.D. (1972), "Developmental Model of the Southern Appalachians," *Geological Society of America Bull.*, Vol. 83, pp. 2735-2760.

Healy, H.G. (1975), *Terraces and Shorelines of Florida*: Florida Geol. Survey M.S. 71.

Knight, C.R., Chicago Natural History Museum.

Kurz, D. (1942), Florida Dunes & Shrubs, Florida Geol. Survey M.S. 22.

McNeil, F.S. (1966), *Pleistocene Shorelines in Florida and Georgia*: USGS PROF. PAPER 221-F, pp. 91-107.

Marsh, O.T. (1966, *Geology of Escambia and Santa Rosa Counties, Western Florida Panhandle*: Florida Geol. Survey Bull. 46.

Moore, R.C. (1949), *Introduction to Historical Geology,* McGraw-Hill Book Co., New York, p. 582.

Murray, G.E. (1961), *Geology of the Atlantic and Gulf Coastal Province of North America*: Harper & Brothers, New York, p. 692.

Puri, H.S. & R.O. Vernon (1964), *Summary of the Geology of Florida and a Guidebook to the Classic Exposures*, Special Publication 5, Florida Bureau of Geology.

Pfleider, G.P. (editor). *Surface Mining*, Am. Inst. of Mining, Metallurgical, & Petroleum Engr. 1058 p.

Sawkins, F.J., G.C. Clement, D.G. Darby & G. Rapp (1974), *The Evolving Earth*, Macmillan Publishing Co., New York, p. 477.

Sullivan, W. (1984), "Geologist add more Pieces to a Global Jigsaw Puzzle," *Discover*, pp. 66-75.

Vernon, R.O. (1943), Florida Mineral Industry, Florida Geol Survey Bull #24.

Watkins, R.O. (1971), "Sea-floor Spreading 196 Crustal Convergence Model for the Southern Appalacians," *Geological Society of America*, abs. with Programs (Southeastern Section), Vol. 3, No. 5, pp. 357-358.

Wolman, M.G., & L.B. Leopold (1957), River Flood Plains: Some observations on their formation, U.S. Geological Survey Professional Paper 282 C.

SUMMARY OF CHAPTER 4

The topography of Florida is strongly influenced by four primary natural forces. These include (1) surface erosion by running water; (2) solution action by groundwater; (3) the work of winds, waves, and ocean currents; and (4) sea level fluctuations. Man's activities are becoming progressively important topographic influences as well.

Florida's central chain of sand hills began as offshore bars and barrier islands during Pleistocene times. Winds and changing sea levels eventually relocated these hills to a more central and landward location. The stability of both sand hills and beach dunes is promoted by their vegetal cover.

Man's surface mining and urban development has promoted extensive surface modifications, not only through direct sculpturing of the surface, but through such correlative effects as erosion and drainage modifications.

The Floridan Plateau is divided into four physiographic divisions. Each division has its own distinct topography and natural features. The divisions are described as (1) Northern Highlands, (2) Central Highlands, (3) Marianna Lowlands, and (4) Coastal Lowlands.

Florida receives the equivalent of 150 billion gallons of rainfall daily. Nearly two-thirds of this is lost to evaporation, 25 percent is expended as stream discharge, and the remainder as underground flow. The surface drainage patterns are controlled by the pattern of hills, terraces, and a cross-peninsular divide.

REVIEW OF CHAPTER 4

Part A

1. The erosion effect of surface runoff is most effective in northern Florida because
   a) this area receives more rain than found in south Florida
   b) the energy levels of the running water are significantly greater here
   c) the soils in the region more readily dissolve in fresh water than further south
   d) the clay content reduces rain water infiltration
2. The Central Florida landscape has been controlled primarily by
   a) the erosive effect of its rivers
   b) limestone solution by groundwater
   c) sand deposits created by sea winds
   d) formation of Pleistocene terraces
3. The factor controlling the division between Florida's two karst topography districts is
   a) thickness of the limestone
   b) thickness of the Hawthorne formation or the lack of this impervious layer
   c) thickness of beach sands
   d) existence of Pleistocene terraces
4. Coastal dunes retrograde naturally due to
   a) encroachment by the sea
   b) blowout process
   c) undermining by waves
   d) both b and c
   e) all the above
5. Vegetation progression on coastal dunes proceeds from ________ on foredunes ________ to on inland dunes.
   a) sea oats, saw palmettos
   b) low shrubs, oak trees
   c) saw palmettos, oak trees
   d) sea oats, oak trees
6. Man's change of the Florida landscape may be attributed to all but which of the following
   a) coastal development
   b) increased use of energy
   c) urbanization
   d) increasing population
   e) all of the above are important factors

7. Match the following geomorphic features with the corresponding *physiographic: division* by placing the corresponding letter in the blank.

____	7.	Trail Ridge	A. Northern Highlands
____	8.	mangrove swamps	B. Central Highlands
____	9.	Tallahassee Hills	C. Coastal Lowlands
____	10.	Holmes Balley Scarp	D. Marianna Lowlands
____	11.	Big Cypress Swamp	
____	12.	Sand Hills region	

Part B

1. What primary natural forces shape the topography of Florida? How does each modify the surface?

2. Why do Florida's highlands resemble dunes found along the coasts?

3. What important features control the drainage patterns of peninsular Florida? How does each feature affect the drainage?

4. Describe the 3 types of rivers found in Florida including the features that distinguish each.

5. What has been the impact of construction of the cross Florida Barge Canal?

6. How do quicksand pits form along some Florida streams? How dangerous is quicksand?

Part C—Map Exercise

Using a map from the appendix draw in the major rivers of Florida, and using a color code indicate what type of river each is.

CHAPTER 5
The Subsurface Processes Molding Florida

We have examined nature's methods of directly molding the Florida landscape. However, nature's subsurface processes also leave their unique footprints on the Florida topography. Lands north of the cross-peninsular divide routinely reflect alteration by subsurface processes. Here, surface sands are relatively thin and often poorly cover the activities below. South of Lake Okeechobee, thick distal deposits effectively mask nature's modification of the underlying strata.

Florida's subsurface geomorphic processes may be divided into two major areas—tectonics and groundwater activity. The latter process is perhaps more spectacular, as associated sinkholes form rapidly, often become lakes, and somewhat routinely devour man's structures and properties. Conversely, Florida has not experienced a large earthquake associated with tectonics in several decades. Surface expression of Florida's tectonics may require centuries before underlying processes are recognized.

As populations expand and urbanization advances into broader areas, man's stimulation of subsurface processes may be expected. Much of the stimulation will be attributed to manipulation of surface and ground waters.

TECTONIC PROCESSES AND THE FLORIDA LANDSCAPE

Two tectonic processes have actively contributed to the modern landscape of Florida. The action of these processes may be considered recent (for the most part) by geologic time standards. The application of compression and tensional forces produces rock deformation and fracture. Bending and deforming of the rocks is termed **folding,** while actual rock breakage results in joints and faults. **Joints** reflect fractures along which no movement or displacement of the rock has occurred; conversely, **faults** represent fractures along which some rock movement has occurred. Weak rocks are more likely to fold and deform, while fracturing is common in brittle rocks. Folding normally occurs at depths where temperature and pressure are high. Faulting is recognized as a near surface phenomena.

Folding Processes

Folding Florida's layered rocks produces wavelike patterns within the strata. Broad fold systems, of which the Paleozoic Peninsular Arch is a part, are associated with the concept of plate tectonics. Small scale and localized deformation are normally associated with **isostasy** or regional loading stresses.

A variety of terms are applied to folded systems. An **anticline** represents an upward folded, hill-like structure. **Syncline** is the name applied to downward folds which create basins or valley-like structures. These folded structures vary in size from a few inches in width and length to dimensions measured in tens of miles. The prefix "geo" is added to describe such earth structures hundreds of miles in length; thus the terms **geoanticline** and **geosyncline**. In Florida, many of the folds may be described as **plunging.** This implies that the fold is not horizontal but dips more steeply at one end than the other. Domes represent small anticlines which are somewhat symmetrical and plunge at both ends. Domes of the Panhandle area contain portions of the state's oil reserves.

Three Cenozoic anticlines still influence the Florida topography of today. They include the Ocala Uplift, the Sanford High, and the Chattahoochee Anticline. All have been severely eroded until their surface expression is difficult to recognize. These uplifts are shown in Figure 5-1.

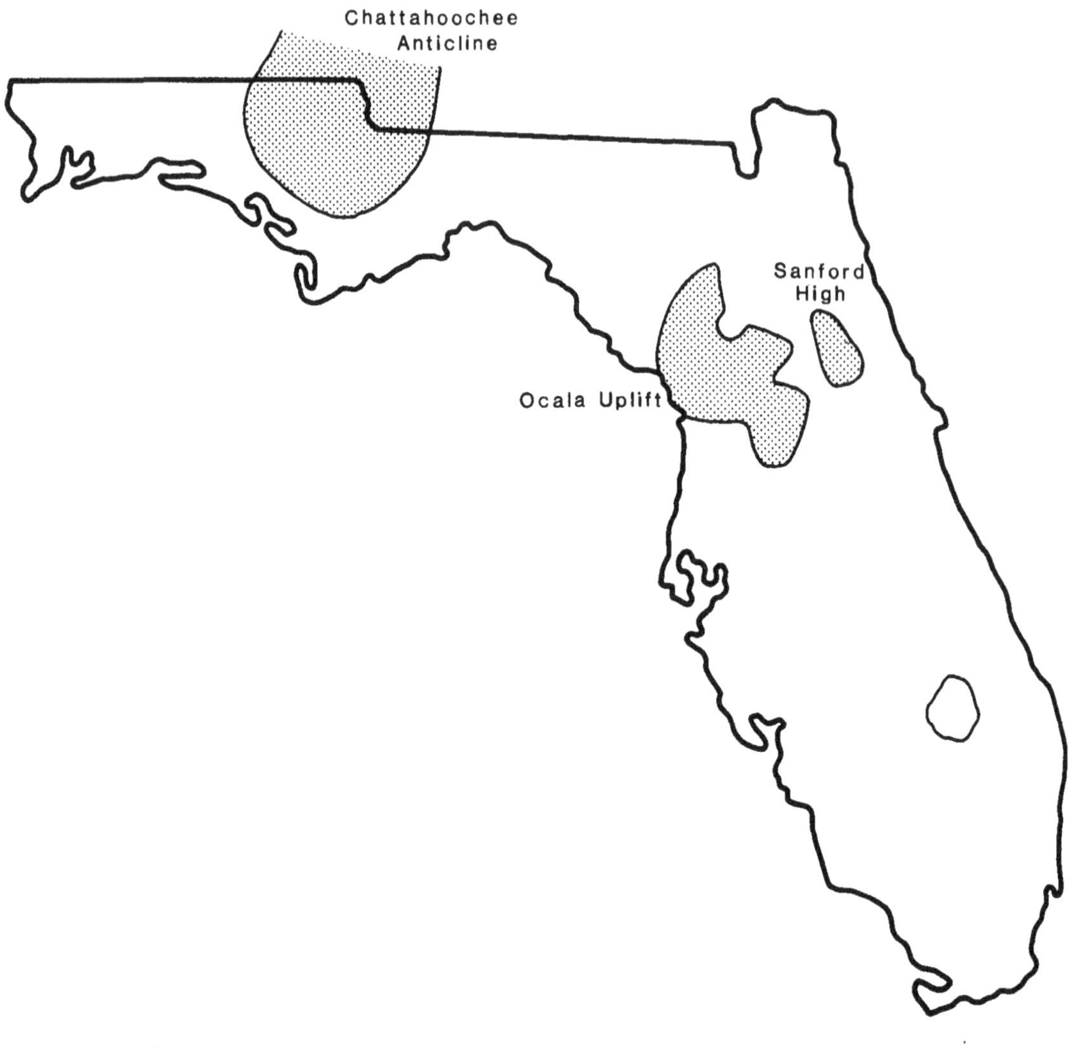

Figure 5-1. Important Cenozoic uplifts affecting the Florida topography.

The Ocala Uplift is a gentle flexure approximately 230 miles long and some 70 miles in width. The uplift trends northwest-southeast with extensive faults. Studies of the uplift's structure indicate it could have resulted only from compressive stresses applied from the west and directed against the Peninsular Arch. The presence of Miocene sands, clays and phosphates (termed the Hawthorn formation) along flanks of the structure suggest its uplift continued over millions of years, although the region was submerged during the middle Miocene.

The Chattahoochee Anticline is a broad flexure extending northward from Florida into Alabama and Georgia. The structure trends northeast-southwest and crests in Jackson, Holmes, and Washington counties.

Both uplifts are primarily responsible for exposure of Eocene limestones within the state and exert a significant influence on modern day karst and sinkhole topography.

The Sanford High exercises a minimal influence on modern Florida's landscape. This high consists of a half dome or closed fold in the vicinity of Sanford, Florida.

The south Florida Embayment existed as an important syncline during the Cretaceous Period. The marine sediments accumulating in this syncline since the Cretaceous have given the Everglades and all south Florida its present form and character.

Fracture Systems

Crustal fractures are prolific over the entire Florida Plateau and no region of the state has escaped this phenomenon. Many of the joints and faults are deeply covered by sediments. Others are near surface features and can be readily recognized from air photos. Throughout Florida, joints tend to be localized features, and while many localized faults are recognized, the major fault systems extend through multiple counties.

An understanding of Florida's fracture systems is promoted by knowing the terminology employed. A **fault plane** is the plane of breakage between two fault blocks. Relative movement of the blocks occurs along this plane. The surface trace of the fault plane is a **fault line**. When one fault block moves vertically with respect to the other, a cliff or steep slope is produced. This is termed a **fault scarp**. In most cases we are unable to determine the exact amount of movement along a fault plane because we cannot find exact or corresponding points on both sides of the fault to indicate where the rocks were originally joined. Thus we have no reference point from which to measure movement. Additionally, many fault planes are curved and adjacent rocks deformed, thus compounding movement study problems.

There are a variety of fault types that occur in Florida. These are summarized in the accompanying chart (Figure 5-2). It should be recognized that few faults include displacement in only simple directions as

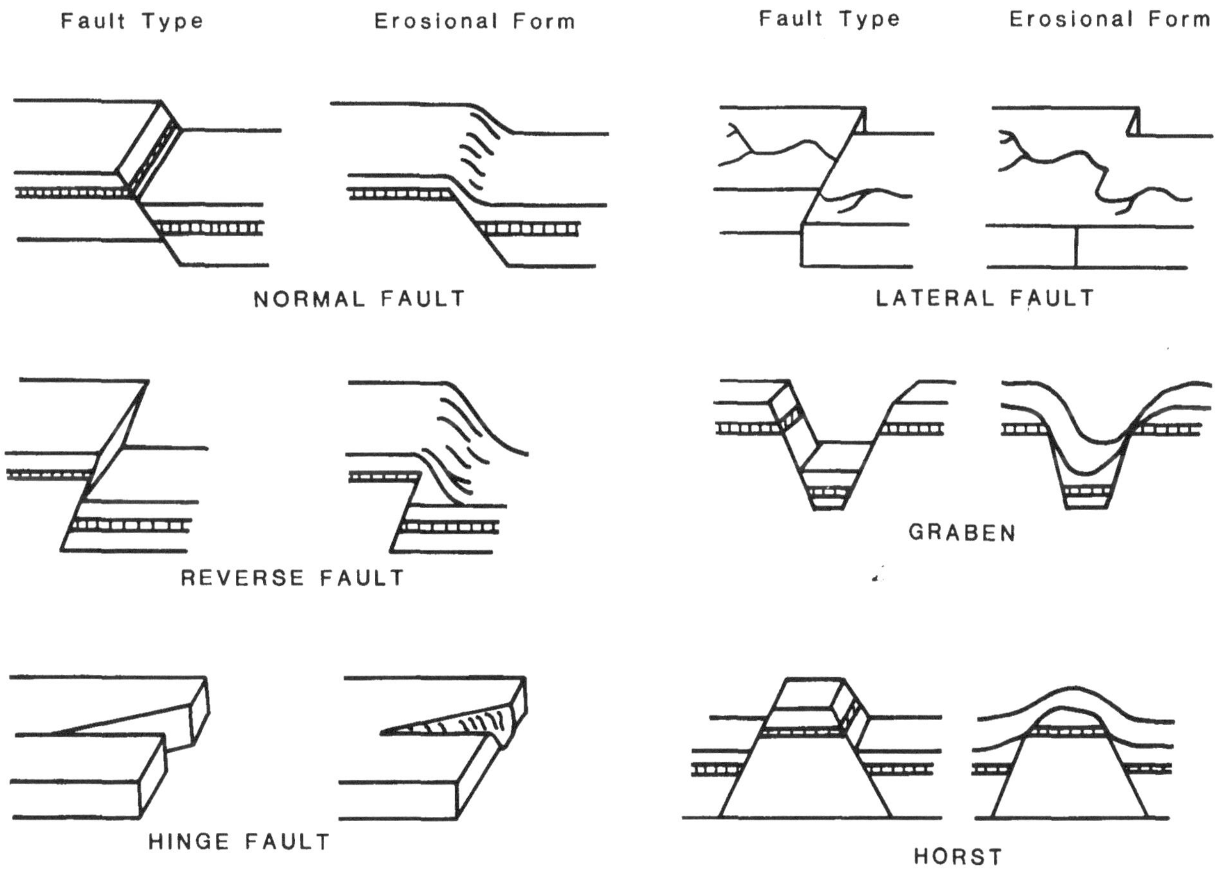

Figure 5-2. Fault types recognized in Florida. The diagrams show how the faults are formed and their eroded surface appearance. In the case of lateral faults, the shift is recognized by offsets in surface features or changes in stream patterns.

suggested in the chart. Most faults include multi-directional movement. However, in describing the faults, the name corresponding to their primary movement direction is generally applied. Initial movement or slippage along Florida's faults has varied from one inch to 50 to 60 feet. Successive movements often occurred over a period of centuries, occasionally resulting in an accumulated displacement of several miles. The slippage along most fault planes is sudden and generally produces strong vibrations or earthquakes in the ground. Florida's recent earthquakes are listed in Table 5-1.

Table 5-1. Florida Earthquake History

Date	Description
November 6, 1978	A tremor with epicenter near Lake City shook northern Florida. Richter Scale reading was 1.8.
January 12, 1978	A tremor originating somewhere in Central Florida caused windows and walls to shake.
December 4, 1975	An earthquake of approximately Richter intensity 3 occurred southwest of Daytona Beaach. Several structures were slightly damaged. The earthquake was attributed to either shifting of the St. Johns River Fault or isostatic uplift along the continental shelf.
October 31, 1900	An earthquake with epicenter between Lake City and Jacksonville produced slight structural damage to residential structures.
January 12, 1879	A tremor felt from Tallahassee to Daytona rocked an area of 25,000 square miles. The eaarthquake registered 6.0 on the Richter Scale and produced some structural damage to buildings..

Fracture patterns throughout the state (Figure 5-3) were first recognized by Vernon (1951), who mapped the patterns from their surface appearance on aerial photographs. Fractures appear as continuous trough-like depressions and ridges marked by significant vegetation changes and soil colorations. Many of the fractures have been verified by geologic core sections, particularly along the cancelled cross-peninsula barge canal. Faults commonly found in central Florida have small horizontal displacements of 10 to 20 feet. Several of Vernon's faults, however, are questionable, as borings have not been able to locate them. This is particularly true for faults in eastern Orange and Osceola Counties.

Throughout much of central Florida the fracture patterns lie parallel to the axis of the Ocala uplift and help in locating the exact crest of this fold. These fractures trend from the northwest to the southeast and represent the major system in the state. A secondary system of fractures trends northeast-southwest and intersects the previous system at irregular intervals.

Regional patterns are grouped irregularly at wide intervals of 20 to 30 miles and form rough rectangular patterns where they intersect. These patterns may be recognized throughout most of the state. Two fracture patterns which differ from the rectangular system have been recognized, however. Radial patterns are recognized in several areas, including both the Everglades and Panhandle. Such radial patterns are also common along the Gulf Coast and generally indicate subsurface salt domes.

Geologists have recognized that the fracture patterns correlate well with features of the peninsula's crystalline basement. To do so, the fractures must either (1) pass downward through several thousand feet of Cretaceous, Tertiary and younger sediments; or (2) result from readjustments and compaction of relatively unconsolidated sediments overlying the stable basement rock.

Modern fractures may be attributed to isostatic rock adjustments resulting from stresses produced during Pleistocene shifting of oceans and to earth tides.

Figure 5-4. Fracture pattern of the northern portion of the Florida Peninsula. (Reprinted by permission of the Bureau of Geology, State of Florida Department of Natural Resources.

The Ocala Uplift

Many of the recent features in central Florida may be associated with the formation of the Ocala Uplift. The compressive forces elevating this anticline were opposed both by tensional stresses and the weight of the rock mass itself. This stress combination produced the fracture patterns accompanying the uplift. The series of fractures parallel to the fold axis are predominantly normal faults. These tensional joints tend to die out with depth, as stretching is greatest toward the outside of the flexure. Several of the faults have formed graben depressions, horst ridges, and flattened the crest of the uplift. Vertical displacements of 20 to 160 feet are recognized. The fault scarps tend to be steeply inclined. Only a minimal indication of horizontal rock displacement is apparent. The extensive fracture patterns are promoted by the tendency of Ocala limestones to favor strain adjustments by step fracturing rather than by bending or flexing. (See Figure 5-4)

Stream patterns and sinkhole alignment commonly parallel fracture systems. Particularly well developed expression of these fractures is shown by their influence on the Oklawaha, Withlacoochee, and Kissimmee river valleys. Each of these rivers shows strongly developed, rectangular trends and sharp angular turns. Most of Florida's larger springs lie in or along these fractures.

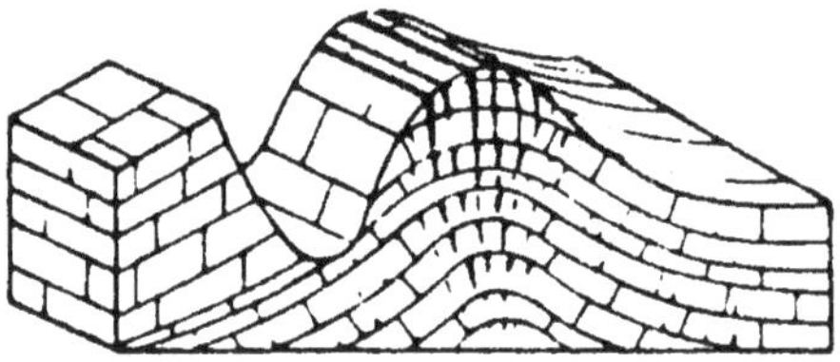

Figure 5-4. How cracks form in an anticline as a result of folding as in the case of the Ocala Uplift. (Reprinted by permission of Jammal and Associates, Inc. and the City of Winter Park.)

Central Florida Tectonics

Geologically recent tectonics have occurred primarily in central Florida. The important tectonic features as described by Vernon (1951) include the Kissimmee Faulted Flexure, the Sanford High and the Osceola Low. The Volusia Graben and the miscellaneous fracture systems of Polk and Lake Counties are also recognizable but are considered of lesser structural importance. None of the faulted features described has received significant investigation.

FLORIDA'S GROUND WATER SYSTEMS

Rain falls somewhere along the Florida Peninsula almost daily. Much of this water is returned to the atmosphere by evaporation and transpiration from plant leaves or needles. The waters which are not lost to the atmosphere eventually return to the sea by overland flow, or they soak into the pore spaces of the soil in a process called **infiltration**. The water from brief showers has little opportunity to infiltrate deeply before it is evaporated by the sun's warmth or taken up by shallow roots. However, water from a heavy and prolonged rainfall tends to infiltrate more deeply and continuously until the aquifer storage space is filled. Surface runoff then begins. Surface runoff is usually minimal in natural or rural areas. It is, however, a problem in urban areas.

The ability of water to infiltrate the soils is directly proportional to the soil's permeability or ease of fluid transmission. Permeability decreases with depth and increasing clay content. Preferential permeability in some horizontal direction may result from progressive increase in soil porosity characteristics, subsurface erosion, or other features.

Florida's infiltrating waters move downward through the relatively porous surface sands until encountering clays, phosphates, hardpans, or other layers with low porosity. Such layers which retard downward water movement are called **aquicludes**. The major aquiclude throughout the state is the Hawthorn Formation described in a previous chapter. The Hawthorn Formation has been breached by erosion in the Marianna Lowlands and along portions of the Ocala Uplift. The aquicludes so retard further downward movement that the water "backs up" and saturates or completely fills the soil pore spaces. The result is a saturated or water storage zone whose upper boundary is called the **water table**. Between the water table and the surface we often find an unsaturated zone in which the pore spaces contain both air and water.

Large areas which are capable of holding and transmitting groundwater are called aquifers. When the water table is free to rise and fall in response to rainfall and discharge variations, the aquifer is termed **non-artesian**. In Florida non-artesian aquifers tend to lie at relatively shallow depths.

Over prolonged periods of time the waters of the saturated zone infiltrate through the aquiclude and into other porous layers below. Much of Florida is underlaid with porous limestone, and it is in this strata that most of the potable aquifer lies. Where the limestones or other porous layers are confined between strata of low permeability, the waters they contain are subjected to greater than atmospheric pressure. The resulting aquifers are termed **artesian.** (Figure 5-5) The waters of wells drilled into artesian aquifers normally rise under hydraulic pressure to a level higher than the overlying aquiclude. The level to which the pressurized waters rise is called the **piezometric level**. In several of Florida's large springs the piezometric level lies above the land surface, and large volumes of water are literally pumped naturally over the surface. Several of these such as Silver and Homosassa Springs deliver over 60 million gallons of water daily.

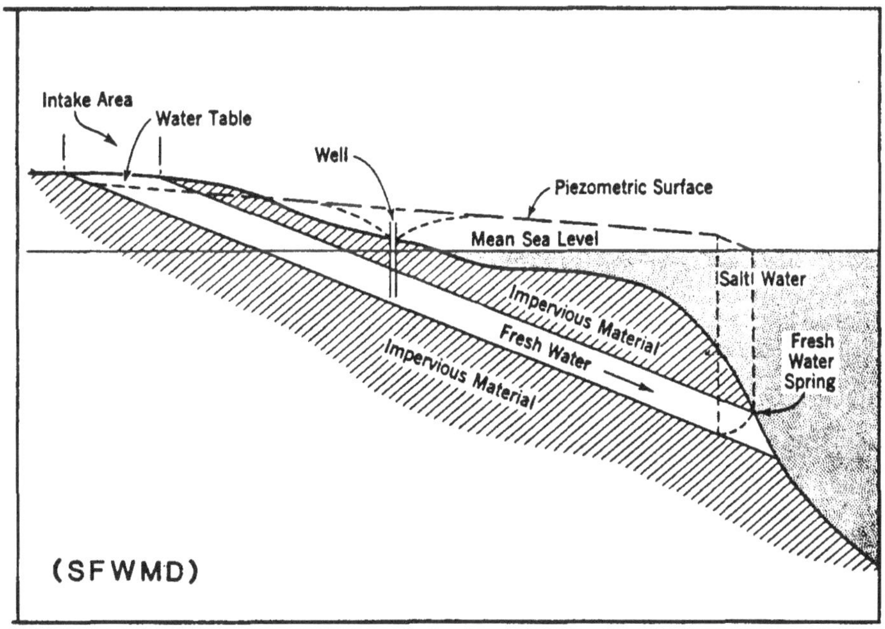

Figure 5-5. In an artesian aquifer, the aquifer or water layer is confined between two impervious aquicludes. The water rises from the well under pressure. (South Florida Water Management District.)

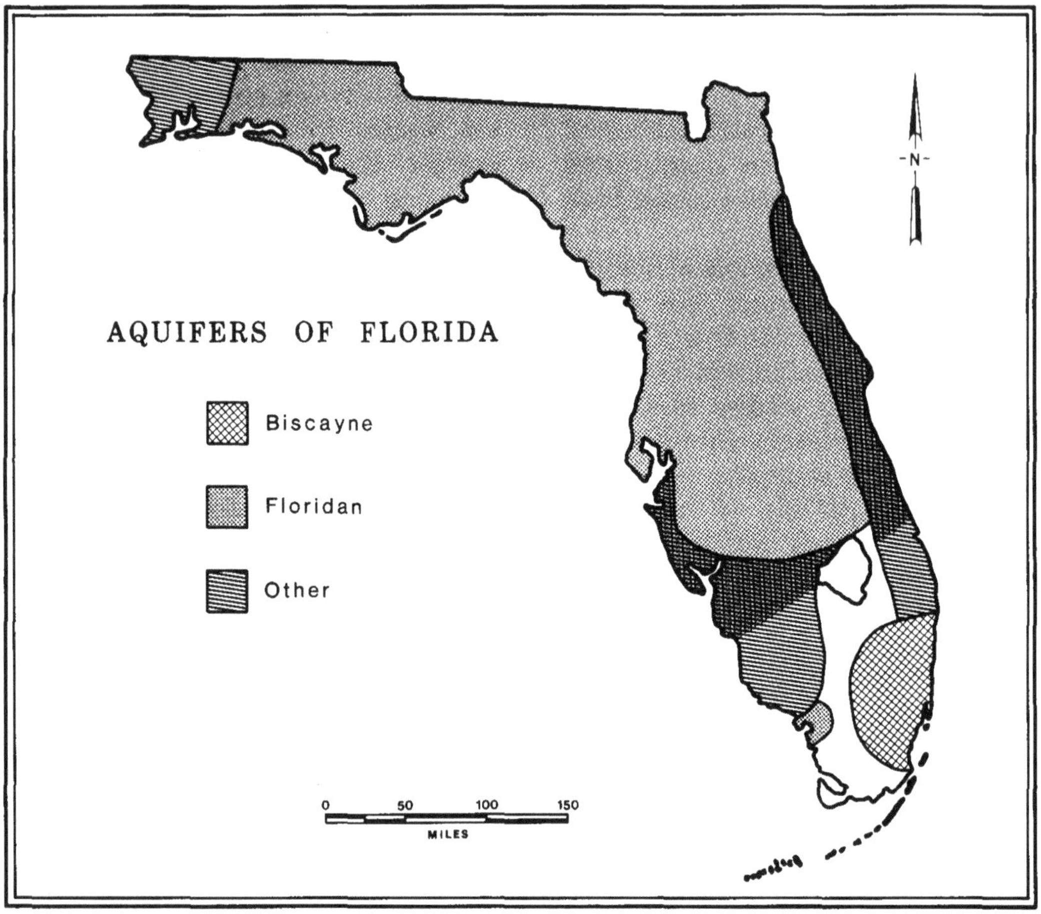

Figure 5-6. Principal aquifers of Florida. (Reprinted by permission of the Bureau of Geology, State of Florida Department of Natural Resources.)

Florida's Aquifers

Florida is blessed with abundant groundwater resources. Four major aquifers are found within the state (Figure 5-6). The Floridan aquifer is the only artesian aquifer, but it provides the state's primary source of water. The state's non-artesian aquifers include the Biscayne aquifer of extreme southern Florida, the sand-and-gravel aquifer of the western Panhandle, and the shallow aquifer of the east coast and south-central region. These aquifers are shown in the accompanying figure.

The Floridan aquifer is contained almost exclusively in the thick limestones which underlie the entire state. In the west coastal areas and the Ocala Uplift, the Floridan aquifer lies near the surface. Throughout central Florida the aquifer lies at 100 to 200 feet beneath the surface. However, as one moves progressively southward, the Ocala and the underlying Avon Park limestones (and the accompanying aquifer) become more deeply buried. Beneath the Everglades this aquifer lies beneath some 1,200 feet of sediments. The Floridan aquifer has a water storage capacity in excess of all five Great Lakes combined.

The non-artesian Biscayne aquifer is the primary source of water for the southeast coastal counties of Dade, Broward, and Palm Beach. This aquifer exists primarily in permeable sands and limestones of Miocene through Pleistocene ages. Beneath the Biscayne aquifer lie impermeable beds of clayey marl. These beds shield the aquifer from the intrusion of salt water found in the limestone strata below. Unfortunately, the Biscayne aquifer lacks lateral protection from the salt water intrusion now progressing inland through near-surface beds. The water of the Biscayne aquifer is considered "hard," as it contains large amounts of dissolved carbonates (200 to 300 ppm) and is stained with dissolved iron.

The sand-and-gravel aquifer lies in extensive erosion deposits left by streams draining from Georgia and Alabama. The aquifer consists of mixed quartz sands and pebbles, chert gravels, and clay stringers. Water of the sand-and-gravel aquifer is relatively pure and is considered "soft," as the level of dissolved carbonates is very low.

The shallow aquifer underlying the Everglades and Big Cypress Swamp is the major water source for the eastern and southwestern coastal counties. The waters of this aquifer vary from "soft" to "hard" depending upon the well location. The variance in water quality correlates directly with the local variations in lithology, thickness, and characteristics of the beds in which the aquifer lies. Where the aquifer occupies Pamlico age sands, the waters are relatively soft; however, in younger beds containing large quantities of shells and limestones the waters are noticeably "hard." Dissolved iron levels are high throughout most of the region.

Non-artesian aquifers lie close to the surface throughout much of the state. Only beneath the area of rolling sand hills are they found at considerably greater depths. In the lowlands and flatwood areas the water table normally varies from 2 to 10 feet beneath the surface. The permeability of near-surface soils and the beds immediately beneath them determine the extent to which the water table conforms to the surface topography. The water table is usually near the surface where impervious beds lie at very shallow depths.

Shallow aquifers show a rapid response to both heavy rainfalls and dry periods. Because local rainfall is the primary source of water, heavy rains quickly raise the level of the water table. During extended dry periods the water table lowers rapidly due to the combined effects of evaporation and transpiration (called evapotranspiration), discharge into streams and lakes, and downward leakage through the aquiclude. During dry periods water table levels tend to stabilize at about 5 to 6 feet beneath the surface. At these depths the water table lies well below most stream beds, evaporation practically ceases, and the water level falls well below the level of shallow root systems. Downward leakage through the aquiclude also declines due to lowering of the water pressure around the impervious beds.

In many areas the non-artesian aquifers occur at two or more levels. The soils above the Hawthorn Formation lie in alternating beds of porous sand (or clayey sand) and less permeable clays or sandy clays. Each porous layer in effect becomes a secondary or semi-artesian aquifer, confined between entrapping layers. Each secondary aquifer exhibits its own piezometric pressure. Studies of the soils surrounding the famous Winter Park sinkhole indicated a semi-artesian aquifer lying 50 to 65 feet beneath the surface. The Hawthorn Formation separated this secondary aquifer from the Floridan aquifer below. The Hawthorn Formation at this point was 90 feet thick.

Localized faults and joints provide conduits or pipes for water flow between secondary aquifers. They also provide direct penetration routes for water movement through the Hawthorn aquiclude. Aquifer water levels are depressed around such fractures, in what are termed depression cones. Similar depression cones are found also surrounding drilled wells. In both situations depression cones result from localized water withdrawal at rates faster than it is replenished (Figure 5-7).

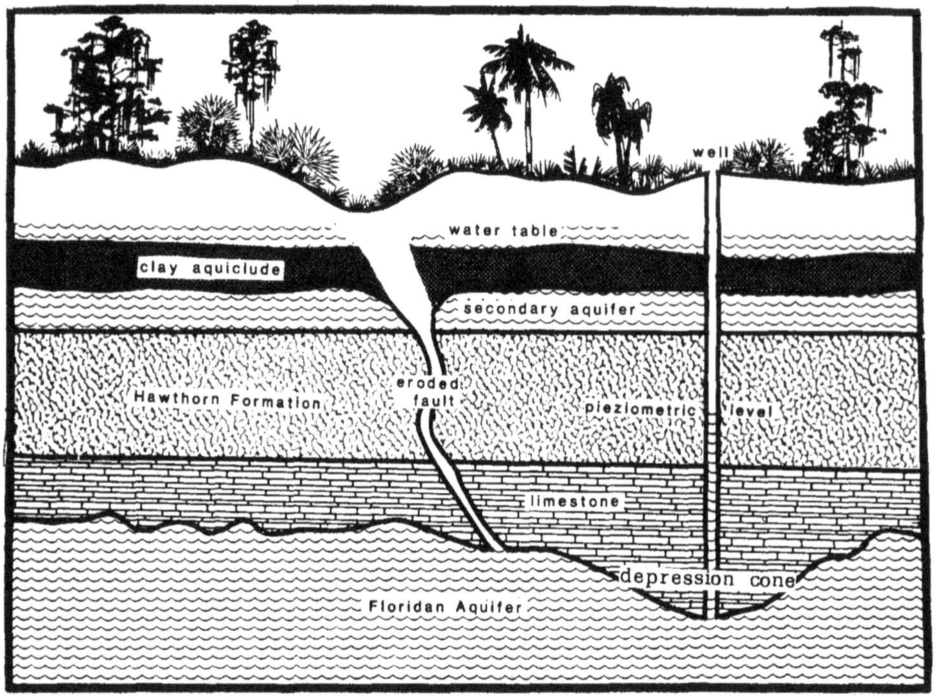

Figure 5-7. Shallow non-artesian aquifers recharge the Floridan Aquifer through a fault. Depression cones are shown in the shallow aquifers adjacent to the fault. Artesian water rises to the piezometric level in a nearby well.

The presence of subsurface fractures, depression cones, and drainage is often easily recognized at the surface. Vegetation which normally depends upon shallow groundwaters becomes stressed as depression cones rob it of normal water supplies. It is not uncommon to see circular or elongated areas of vegetation so stressed. Trees which depend on tap roots to reach the water table are particularly affected as depression cones expand.

Over the years groundwaters become progressively diverted into the local fractures. The increased water flow through these fractures erodes the loose materials and carries them away. Erosion creates enlarged subterranean piping systems, which in turn are capable of even greater ground water movement.

Processes within the Floridan Aquifer

Groundwaters flowing through the Hawthorn aquiclude spread throughout the limestones of the underlying Floridan aquifer. This process, where water is added to the aquifer, is termed aquifer recharge. Recharge areas are the primary entry points through which an artesian aquifer is recharged. The water movement in the limestones tends to follow fractures and bedding planes in the rock. These features provide a relatively free avenue for water circulation.

The groundwaters of Florida tend to be acidic due to the absorption of carbon dioxide which readily reacts to form carbonic acid. Carbon dioxide is provided by bacteria action on organic surface materials (i.e., dead leaves, grasses, etc.). Modern "acid rains" also contain dissolved sulfuric(ous) and nitric(ous) acids as well.

Solution and removal of the limestones by the mildly acidic waters is most effective and rapid along the fracture and bedding planes, particularly at their intersections. Over the centuries these water flow paths expand into extensive and interconnected cavern systems. Water flows through these pipes and caverns in much the same manner as through man's surface drainage and sewer systems. Depth of the cavern systems tends to increase progressively with distance from the recharge sites. The depth of solution activity is controlled by existing base level at the time of solution activity.

Water moves much more slowly in the solid but porous rock. Solution activity is slowed correspondingly. Small primary pore spaces slowly enlarge, coalesce, and eventually a fine textured, honeycomb, or spongiform rock results. The shells and remains of marine invertebrates in the limestone dissolve rapidly, leaving behind pockets and holes. The resulting limestones resemble Swiss cheese in appearance (See Figure 4-3).

Today we are seeing the proliferation of urban drainage wells to divert storm water runoff directly from streets into the aquifer. The runoff waters are acidic and contain high CO_2 contents. These waters are introduced locally at greater than normal infiltration pressures. The combination of increased pressure and acidity may be expected to increase limestone erosion. This concept, however, has not been substantiated by conclusive research.

Piezometric maps have been prepared by the Florida Department of Natural Resources to determine areas of recharge, **discharge** (where groundwater returns to the surface as a spring or surface stream), and direction and rate of ground water movement. Groundwater moves from high pressure to low pressure zones. Piezometric mounds (referred to as "highs") usually indicate areas of recharge to the aquifer. Piezometric depressions or troughs (referred to as "lows") usually indicate areas of discharge from the aquifer. Recharge and discharge may occur anywhere between the high to the low where hydrogeologic conditions are favorable. Thus, there is no single point of recharge or of discharge. Aquifer recharge is especially active in areas where the Hawthorn Formation was never deposited, has been stripped away, or is relatively thin.

Currently the piezometric surface of the Floridan aquifer ranges from about 40 to 50 feet below existing ground surfaces. This level fluctuates in accordance with cyclic rainfall differences. The normal fluctuation range is about 10 feet. Overall, however, the piezometric level has been declining since 1930 at a rate of about 0.4 feet per year. The decline is attributed to lowered rainfalls and increased urbanization. Man's construction of impervious roofs, streets, and parking lots increases surface runoff and decreases the amount of recharge water available. Pumping of groundwater for urban activities also lowers aquifer levels and pressures.

Four major recharge areas are found throughout the state. The largest is generally referred to as the "Polk High" and includes the Green Swamp of Polk County. Smaller recharge sites include the Pasco High (Pasco County) and highs located in Liberty and Clay Counties.

Waters recharging the Floridan aquifer are mildly acidic. Chemical reaction between acidic water and overlying sands and clays is negligible. Thus, the water's ability to chemically dissolve the underlying limestones remains high. As the waters flow away from the recharge areas, chemical reactions gradually increase the mineral content. Eventually the waters become saturated. High rates of water movement contribute to increased solution of the limestone with only minimal increase in mineral content levels. Mineral content level is unreliable as a recharge area indicator, however, as water acidity varies significantly with locality.

Soil Subsidence and Sinkhole Formation Processes

Subsidence is a term used to indicate both gradual and rapid settling or depressing of the ground surface due to lack of sub-surface support. Subsidence can occur naturally or be man-induced. Sinkholes are a specialized case of subsidence in which the surface collapses rapidly due to changes in geologic or groundwater conditions. Three distinct types of subsidence are recognized in Florida. The first reflects marshland subsidence and is primarily a surface phenomenon. The other two of these—collapse sinkholes and ravelled sinkholes—form rapidly and are related to conditions in the subsurface limestones.

Marshland Subsidence

Florida's marshes occur on peat and muck soils. Vegetation in these marsh areas was adaptable to varying periods of drought and high, standing water. However, man's deliberate drainage of the marshlands for agricultural and flood control purposes has resulted in a permanent lowering of the water table. The existing marsh plants were unable to adapt to this dry condition and eventually perished. New plants invading the areas did not contribute to the maintenance of organic soils.

Lowering of the water table has thus adversely affected the soil character. The levels of these organic soils has lowered through (1) compaction in which soil particles move closer to one another as the water separating the particles is removed; (2) wind erosion of loose dry soils; and (3) slow oxidation of the dry soil above the water table. Studies by the Everglades Experiment Station show an average subsidence of 1.5 feet during the first 10 years after reclamation of flooded lands. Subsidence continues at the rate of about one inch per year thereafter. Subsidence rates may be increased by muck fires which burn deeply into the peat and muck, destroying the soils completely down to the water level in many instances.

Ultimately, over a million acres of south Florida land is expected to reflect some degree of marshland subsidence as reclamation continues. In many areas subsidence-related problems have produced adverse effects on storm drainage, causing flooding and loss of cultivatable soils.

Collapse Sinkholes

In northern and western coastal areas of Florida the prevailing sinkhole formation mechanism is one of direct collapse of the limestone cavern roof (Figure 5-8). Such failures occur where the limestone is exposed or lies less than about 30 feet below the surface. The usual surface expression in such sinks is initially a dish-shaped depression, with unconsolidated overburden filling the limestone cavity below. Where larger caverns exist, roof collapse often results in near vertical sinks, possibly with exposed rock faces. However, progressive erosion and soil slippage may, in time, alter the sink to more of a funnel shape.

RAVELLING COLLAPSE*

Ravelling is the common subsidence mechanism or sinkhole formation process throughout most of Florida. Ravelling is typical where the overburden is less than approximately 150 to 200 feet in thickness. In the ravelling process, unconsolidated materials (clay, silts, and/or sands) filter downward into voids in the underlying limestone. This movement is promoted by water infiltrating downward as loose material enters the limestone voids (where it may either accumulate or be removed by flushing); a cavity in the overburden develops. The cavity is usually funnel-shaped, with its small opening connected to the limestone opening. The overburden cavity normally enlarges over a period of many years. Where the thickness is shallow (usually less than 20 feet), surface subsidence is common. The usual expression is initially a bowl-shaped sinkhole. Where overburden thickness is greater, the ravelling process may continue until the soil's shear strength can no longer support the arch of overburden; collapse then occurs (Figure 5-9). Allen (1976) recognized that the thicker the overburden, the greater the potential for the information of a large sinkhole.

* The discussionn of ravelling collapse and the subsequent section concerning sinkhole activity in and around Winter Park is reprinted from *The Winter Park Sinkhole Report* prepared by Jammal & Associates, Incc. for the City of Winter Park.

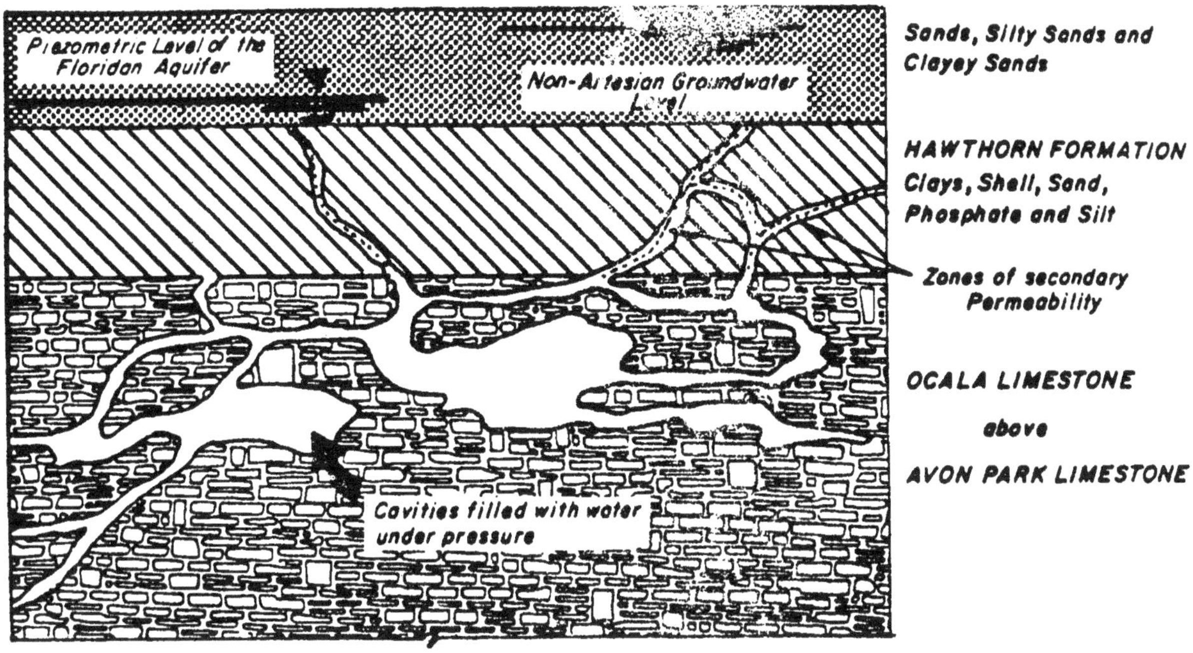

Condition Before Collapse

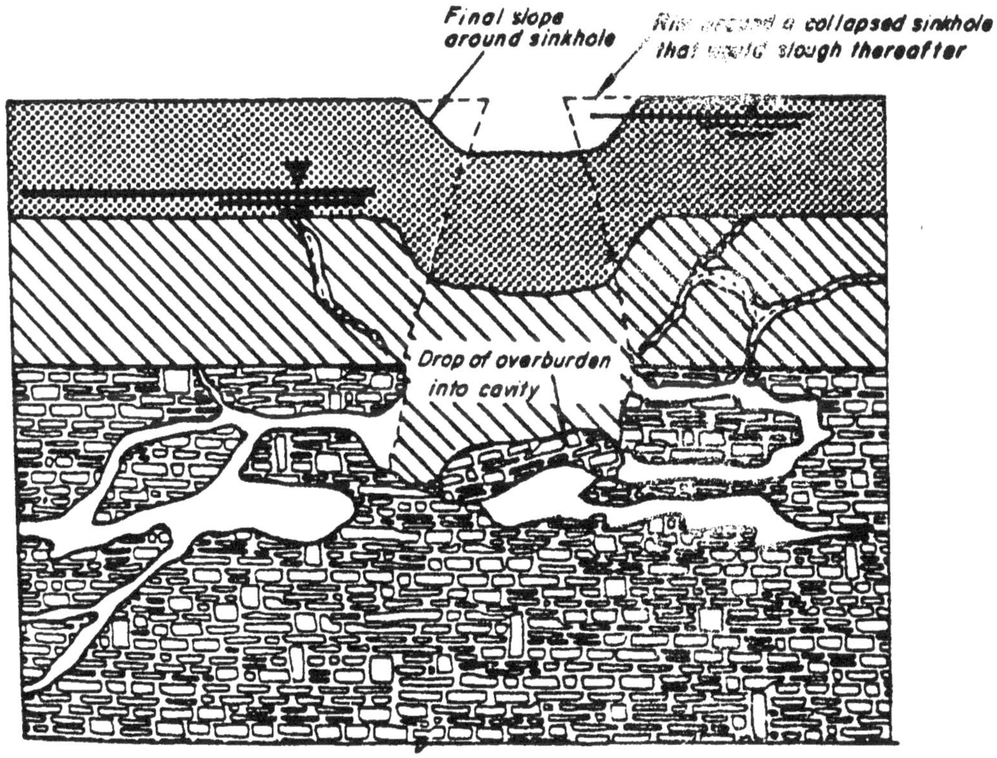

Condition After Collapse

Figure 5-8. Illustration of limestone cavity collapse. (Reprinted by permission of Jammal & Associates, Inc. and the City of Winter Park.)

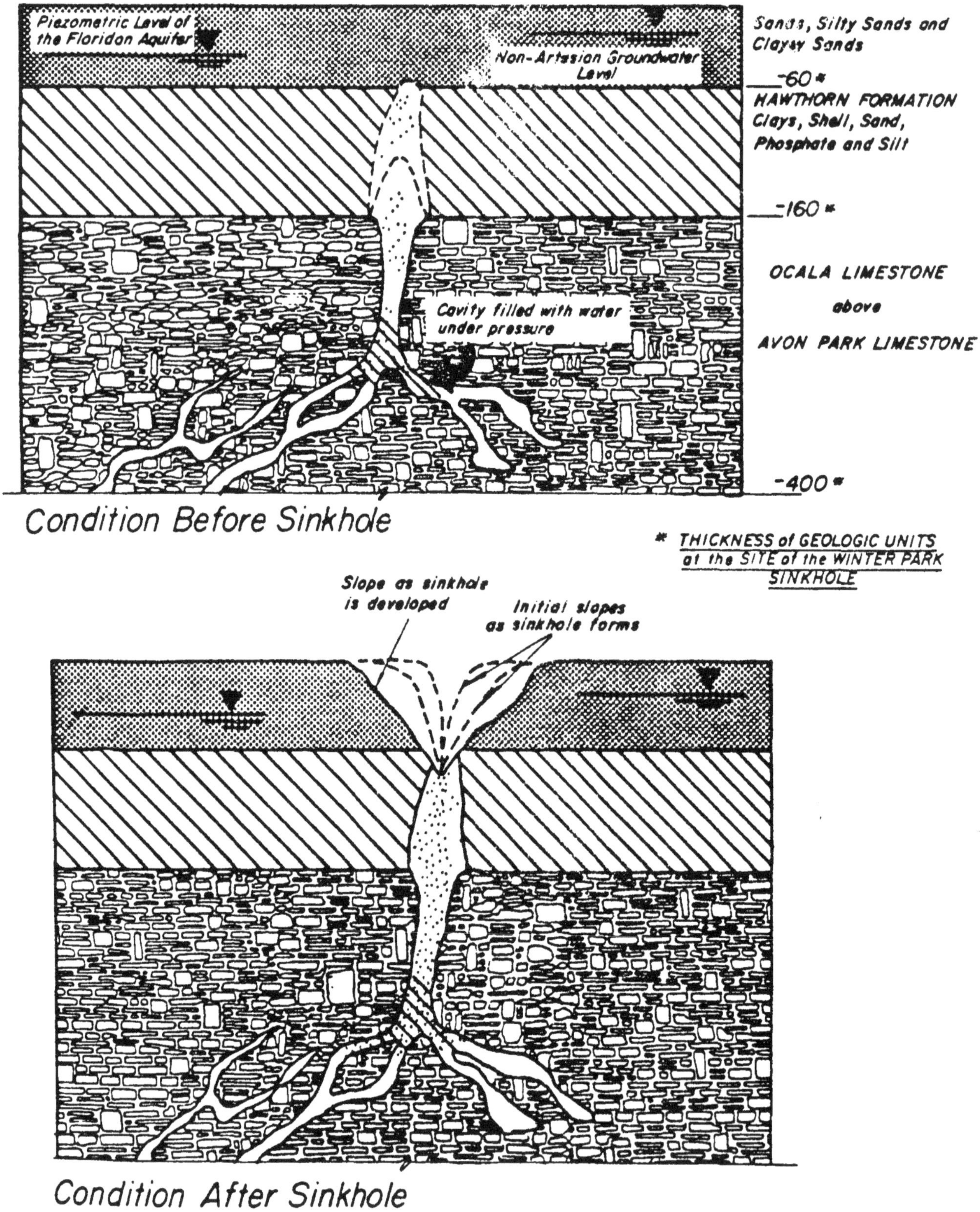

Figure 5-9. Illustration of ravelling collapse. (Reprinted by permission of Jammal and Associates, Inc., and the City of Winter Park.)

Sims (1981) stressed the importance of water table and piezometric levels to sinkhole collapse. High water tables and piezometric levels exert upward buoyancy forces in the soil but tend to weaken soil's shear strength. The high buoyancy forces provide much of the roof support. Dropping water pressures (such as occurs with drought or excessive pumping) lower the buoyancy forces. If the accompanying shear strength increase in the overburden is insufficient to compensate for the loss of buoyancy, cavity roof collapse occurs. Conversely, rapid changes from dry to wet soil conditions promote cavity roof spalling and soil erosion into the limestone opening.

Subsidence (including sinkholes) is common in areas where:

1. Shallow, flat-lying aquifers encompass large areas
2. Water levels fluctuate seasonally
3. Bedding is Pliocene age or younger
4. Clay interbedding is significant
5. Fine-grained aquifers persist
6. Well-developed cavernous zones are common
7. The overburden is less than 500 feet in thickness, and
8. The piezometric surface lies well below the water table.

According to Allen (1976), none of the above features is mandatory, but the more that are present, the greater the subsidence susceptibility. Comparatively, areas of poor subsidence susceptibility contain:

1. Significant cementation of the intergranular voids between soil particles
2. Preconsolidation due to previous deep burial, and
3. Any condition in opposition to those promoting subsidence.

Structural folding also exerts a significant influence on subsidence susceptibility. Subsidence is more likely to occur along the flanks of broad anticlines with flat crest, dip angles of less than 25 degrees, and fractured by tensional faults. Narrow or tightly folded structures with dip angles more than 25 degrees and broken by compression faults seldom have sinkholes. The strata of tight folds are often compressed, have lower porosity and closed fractures than are found in broad anticlines (with tension opened faults).

The major sinkhole-prone areas of the state lie in the Ocala uplift. Windham (1981) indicates that those areas with little subsidence susceptibility probably have:

1. Incomplete flushing of salt water from the aquifer
2. Poorly developed and interconnected caverns
3. Near-surface piezometric levels, and
4. Thick unconsolidated overburdens.

Vernon (1951) indicates that the majority of Florida's natural lakes, sinks, depressions, and ponds are the result of subsidence. The features range from small pits a few feet in diameter to large depressions several miles in size. Most sinks are essentially round. Water normally fills such basins to the elevation of local water levels; where the basin rim lies below the piezometric surface, overflow is common. Lake Alachua, which filled a large subsidence depression for 22 years, was eight miles by four miles.

In the northern areas, cone-shaped sinks (termed ponors) with rocky bottoms and walls are seen. Most are dry or only partially filled with water. Dismal Sink, southeast of Tallahassee, is such a ponor. It is 60 feet in diameter, 175 feet deep and contains water 75 feet deep.

Broadly developed, sediment-filled sinks are referred to as "prairies." The prairies are often marshes. Perhaps the most well-known of these is a prairie area between Gainesville and Ocala, where a coalescence of prairies encompasses a circular area about 10 miles in diameter.

Vernon indicates that other sinks are actually vertical tubes, only a few inches in diameter in some cases, which may extend over 100 feet beneath the surface. These are "natural wells" or "solution pipes," which generally result from localized or pressurized water forced vertically through fractured strata or overburden.

Sinkhole Activity in Winter Park and Surrounding Areas

Sinkhole activity studies in the Winter Park area indicate subsidence is attributable to at least five important factors. These factors include:

1. Fault patterns
2. Surface and subsurface drainage
3. High water table and piezometric level fluctuations
4. Soil permeability and its impact on recharge, and
5. Water hardness.

The Winter Park and Maitland city limits encompass an area which is the apparent terminus of three faults. The faults extend near Lakes Maitland and Osceola, as shown in Figure 5-10.

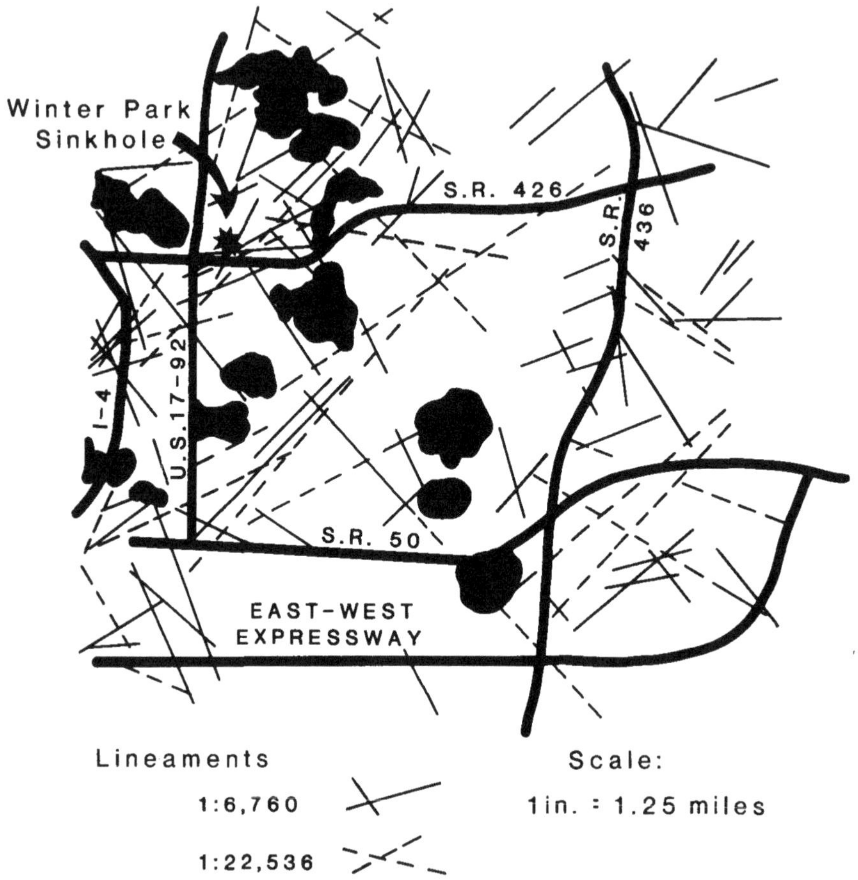

Figure 5-10. Fractures and lineaments in the area of Winter Park. (Reprinted by permission of Florida Department of Transportation.)

Winter Park lies in a topographic basin into which stormwater runoff is diverted naturally. The total area is dominated by well-drained, sandy soils which promote aquifer recharge values of 10 to 21 inches of water annually. The combination of surface topography, subsurface drainage, and porous soils promotes rainwater flow into the three faults. It is expected that solution of underlying limestone begins in and near the faults and is probably greatest where the infiltrating water are most acidic (e.g., in drainage wells).

Well studies within a mile of the Winter Park sinkhole (and within the faulted zone) found abundant evidence of cavity and honeycomb structures in the Avon Park formation.

Water hardness in the vicinity of Winter Park is approximately 60 parts carbonate per million and increases progressively south, east, and west of the area (Florida DNR Map #3). Such changes in hardness indicate the extreme probability of ongoing cavern formation in and about Winter Park.

Aerial and map studies of Orange and Seminole Counties indicate that sinkholes do not appear along the whole length of the faults mentioned. Several explanations for this are possible; however, five are presented:

1. Neither basin drainage nor lake bottom infiltration direct water toward the faults along their entire length; thus, sinks occur commonly near their infiltration point.
2. Water entering a fault proceeds to progressively greater depths as it flows along the fault away from its entry point; the greater the cavern depth the less likely ravelling or collapse is to Occur.
3. Soil qualities and their capacity for aquifer recharge vary along the fault paths, creating nonuniform infiltration and solution rates.
4. Variation in piezometric levels (and hence buoyancy forces) within the fault zones.
5. There is a combination of the preceding factors.

GROUND WATER POLLUTION

Florida is unique among the states in its nearly total reliance upon groundwater for its water needs. Some 92 percent of Florida's population occupying peninsular Florida obtain their drinking water from an aquifer source. This water represents over four billion gallons per day. Much of the remainder of the population living in extreme northern Florida receives its water supplies from rivers originating north of the state. The state's 7500 natural lakes are too shallow to provide continuous potable water.

Thus conservation and protection of Florida's groundwater resources is of major importance. State environmental organizations have recently provided information to the public, which emphasizes the magnitude of the groundwater pollution problem. Their surveys found that over 700,000 tons of hazardous industrial wastes are produced in the five counties of Dade, Broward, Duval, Hillsborough, and Pinellas. These counties contain 54 percent of the state's population. Most of the wastes were found to fall within the definition of hazardous wastes as prescribed by Public Law 94-580, posing "a substantial present or potential hazard to human health or the environment when improperly treated, stored, transported, or disposed of, or otherwise managed." The investigators found that the common methods of disposal include: (1) 62 percent discharge to sewers; (2) 32 percent onsite disposal; (3) 3 percent removal by private haulers (destination unknown); (4) 1 percent landfill; and (5) dredged burial and other methods.

Most of the wastes described are disposed of in such a manner, causing potential contamination of the state's aquifers. The porous and permeable sands covering the state absorb liquids (including contaminants) in a spongelike manner. These contaminants do not all remain in the shallow aquifers but are capable of movement into the Floridan aquifer where it lies close to the surface.

Recent samples of Dade and Broward Counties drinking waters showed small quantities of lead, mercury, arsenic, and silver produced by electroplating and battery manufacturing industries. Concurrently, Duval and Hillsborough Counties experienced chromium, lead, and mercury contamination of their waters (both in groundwater and in spring-fed streams). Potentially dangerous levels of heavy metals were also found in Pinellas County, where the wastes were traced to local smelting, battery, varnish, and paint industries. Comparatively, the water quality in Orange County (with Orlando's industrial and tourist activities) was found to be relatively pollutant free.

The classic case of groundwater pollution problems may be found in the Miami environs. Miami and Dade county receive their drinking water from the Biscayne aquifer. Early wells used by the various municipalities lay near the coast, but as water demands increased these wells became contaminated by intruding sea waters (see following section). The contamination forced the closing of contaminated wells and the progressive drilling of additional wells further inland.

Dade County, however, had created a series of landfills in the western portion of the county. Over the years, containers of industrial wastes buried in these landfills rusted and decomposed. Their contents very slowly spread through the aquifer. The spread of pollutants has been slow, but it has now reached the water of the county wells. Miami's drinking water is now among the most chemically contaminated of any city in the nation.

The Davie Landfill in adjoining Broward County has been used as a site for oil and grease sludge disposal for a number of years. Recent shallow well studies adjacent to the site found high organic carbon and nitrogen levels as well as ammonia and phosphorus, all indicative of sludge pond leaching.

Similar problems are also now being encountered in somewhat diverse situations. Agricultural pesticides sprayed on vegetables and citrus have been found leaching into the soils, affecting the shallow aquifers. Contamination of the Floridan aquifer is also occurring in urban areas where drainage wells introduce stormwater runoff directly into the limestone systems.

Florida's sandy soils do have the ability to cleanse and purify wastewater of bacteria and treated sewage waste. At Carpenter's Home in Polk County, 25,000 gallons of treated waste per day are used to irrigate coastal Bermuda grass plots. Recharge waters leaching to the shallow aquifer from this irrigation project have been drinkable. Similar recycling of "grey water" from sink and tub drains in Longwood (an Orlando suburb) has proved an innovative method of clean water recharge and reduction of sewer line usage.

SALT WATER INTRUSION

For many years we have assumed that unlike murky and undrinkable surface waters, our groundwaters were safe from pollution. The purity of these groundwaters was shielded by a protective basement and by surface sands which filtered and removed contaminants. But we have discovered a silent saline intruder has been progressively stealing into our aquifers. Florida's problems with salt-water intrusion stem primarily from a combination of permeable limestone aquifers, periodic droughts, extensive coastal areas, and the water needs of the populations invading our beaches (Figure 5-11).

Saline water, as defined by the U.S. Geological Survey, is water which contains more than 1000 milligrams (1 gm) of dissolved solids per liter. Highly saline waters are recognized as containing 10 times this volume of solids.

It is recognized that the Floridan Peninsula is surrounded by saline ocean waters. However, the peninsula is also underlain by salty waters occurring at a variety of depths. These waters are termed **connate waters** and are remnants of oceanic waters trapped between the sediment grains deposited in prehistoric seas. Such connate waters are often more than three times as salty as normal ocean water. The saline waters are more dense than fresh water because of their higher dissolved mineral content and thus tend to lie at greater depths than the lighter fresh waters. The equilibrium level of the saline waters fluctuates with changing conditions in the fresh water aquifers.

The Floridan aquifer is a dynamic system. Each foot of fresh water head pressure (in an aquifer above sea level) will depress seawater 40 feet. The reverse is also true. The saltwater level will rise 40 feet for each one foot drop in the water table. In nature, saltwater moves readily through the permeable layers beneath most shallow aquifers. It also occupies the lower levels of the limestones containing the Floridan artesian aquifer, and moves easily through fractures and cavern systems. Under natural conditions, the saltwater intrudes upward during periods of drought in areas where the piezometric pressure declines. Conversely, the saltwaters are forced downward during periods of heavy rainfall.

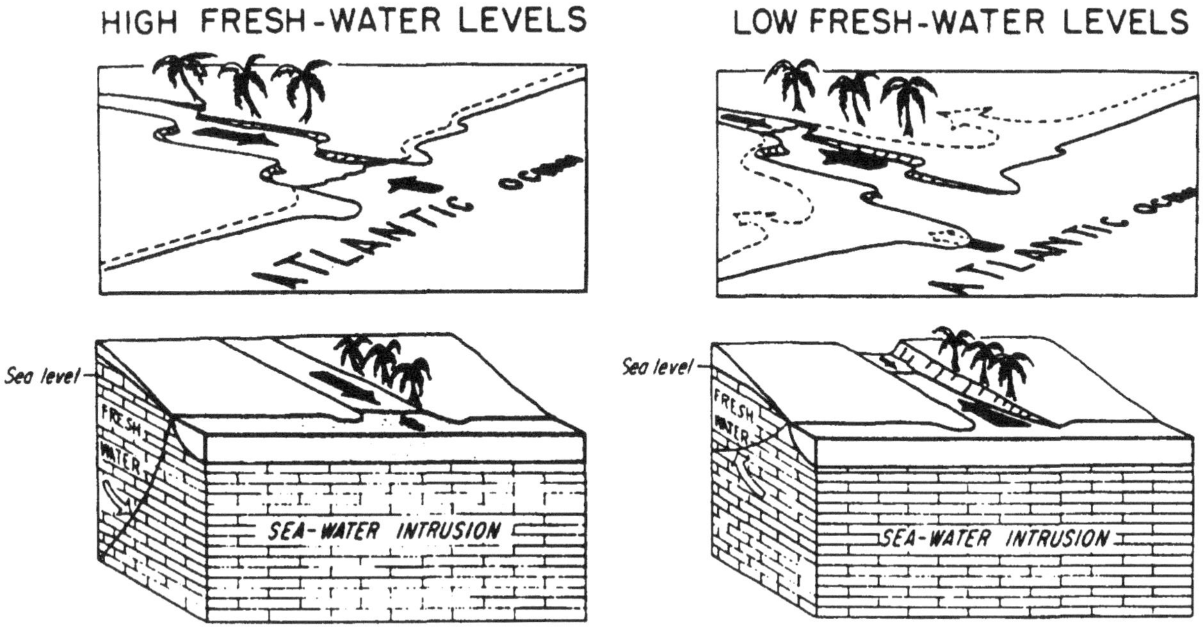

Figure 5-11. Salt water tends to move inland unless balanced by fresh water levels. When fresh water is abundant, stream flow and water levels are high and sea water is forced to remain near the coast. During drought or periods of excessive aquifer pumping, the salty water moves up tidal streams whose flow has been diminished or inland beneath the fresh groundwater.

During the record-breaking drought of 1981 the water levels in many of Florida's lakes dropped so low that shorelines receded hundreds of yards. The recharge potential of such lakes declined correspondingly. Giant Lake Okeechobee, the "heart" of south Florida's recharge system, dropped so low that it no longer provided water for the canal systems radiating from the lake. The lake and its canal systems provide primary recharge for the Biscayne aquifer. As stream flows declined in south Florida, the ocean waters moved inland along the tidal stream channels. Here, their saline waters infiltrated downward as well as intruded laterally along subsurface paths. Water levels in coastal areas of the St. Johns River Water Management District dropped as much as 10 to 12 feet during this drought. This drop in fresh water head allowed rise of saltwater levels in the aquifer of 400 to 480 feet.

Man's activities have created a new dimension in saltwater intrusion. His impact on this can be seen in several ways. As we have previously described, the Biscayne aquifer is underlain by an impervious layer which protects the aquifer from upwelling of salt waters from below. However, this aquifer has no lateral protection from seawater intrusion. Drainage of the Everglades, which began about 1910, has resulted in a progressive lowering of the aquifer's pressure head. This pressure reduction has disturbed the equilibrium of salt and fresh water forces. The balance was tipped in favor of the pressurized ocean waters which advanced inland. Scientists predicted the seawaters would intrude until an equilibrium point was reached at the average 2.5 foot water table level. The saltwater advance enveloped the water supply wells of Miami and West Palm Beach but slowed some eight miles seaward of the predicted equilibrium position by 1955.

Figure 5-12. Controlled canals prevent salt water intrusion and replenish the area ground water levels by influencing fresh water flow into the area. (Reprinted by permission of United States Geological Survey and Florida Bureau of Geology.)

Figure 5-13. Control dam near Miami maintains high fresh water levels near the coast. (Reprinted by permission of South Florida Water Management District.)

The pumping of groundwater has also induced a reduction water head pressure, with results similar to those of surface drainage. Improper well construction, close well spacing, and excess withdrawal rates result in deep depression cones in the water levels surrounding the wells. Two saltwater intrusion phenomena may accompany the formation of depression cones. These phenomena are termed lateral intrusions and upconing. Lateral intrusion occurs when the seawater enters laterally due to reduction of pressure head. Upconing occurs when saline waters below the well rise due to pressure reduction along the fresh water-saltwater interface.

A common practice by many well drillers has been to penetrate the aquifer deeper than necessary. This was an attempt to obtain high water purity and increased flow rates. The result was often direct contamination of the fresh upper zones of fresh water by deep-seated saline waters.

Three major methods are recognized in which coastal saltwater intrusion may be controlled. These include the construction of tidal gates and broadcasting of treated effluent and grey water. Where drainage channels carry surplus waters to the ocean, danger of seawater entry into the channels during high tide and drought exists (Figures 5-12 and 5-13). Lining the channel bottoms near the ocean and installation of tidal gates to permit stream discharge into the ocean while preventing inland flow of seawater has been successful in the Miami area. The spreading or broadcasting of secondarily treated waste from sewage treatment plants and grey water allows the otherwise wasted waters to infiltrate and reestablish former aquifer levels.

REFERENCES

Allen, D.R. (1976), "Subsidence Susceptibility, Methods for Appraisal," *International Association of Hydrological Sciences, Land Subsidence Syjnposium,* Lithocrafters, Ann Arbor, MI, Publication 121.

Applin, P.L. and E.R. Applin (1944), "Regional Subsurface Stratigraphy and Structure of Florida and Southern Georgia," *Am. Assoc. Petroleum Geol. Bull. 28*, pp. 1673-1753.

Back, W. (1963), "Preliminary Results of a Study of CaC03 Saturation of Ground Water in Central Florida," *International Assoc. Sci. Hydrology Bul.* Vol. 8, No. 3, pp. 43-51.

Barraclough, J.T. (1962), *Ground Water Resources of Seminole County Florida,* Florida Geological Survey, R.I. 27.

Hem, J.D. (1959), *Study and Interpretation of the Chemical Characteristics of Natural Water,* USGS Water Supply Paper 1473, 269 pp.

Jammal & Associates, Inc. (1982), *The Winter Park Sinkhole,* a Report for the City of Winter Park, Florida.

Knochenmus, D.D. and M.E. Beard (1971), *Evaluation of Quantity and Quality of Water Resources of Volusia County, Florida,* Florida Geol. Survey R.E. 57.

Lichtler, W.F., W. Anderson, and B.F. Joyner (1968), *Water Resources of Orange County, Florida,* Florida Geol. Survey R.E. 50.

Musgrove, R.H., J.T. Barraclough, and R.G. Grantham, *Water Resources of Escambia and Santa Rosa Counties, Florida,* Florida Geol. Survey R.I. 40.

NOAA (1979), *Earthquake History of the United States,* Washington, D.C.

Pressler, E.D. (1947), Geology and Occurrence of Oil in Florida, *Am Assoc. Petroleum Geologist Bull.,* Vol. 31, pp. 1951-1862.

Pride, R.W., F.W. Mayer, and R.N. Cherry (1966), *Hydrology of Green Swamp Area in Central Florida,* Florida Geol. Survey R.E. 42.

Puri, H.S. (1957), *Stritigraphy and Zonation of the Ocala Group,* Florida Geol. Survey Bull. 38.

Scott, T. (1981), Personal Communication, Florida Geologic Survey.

Sims, M. (1981), Personal Communication, Jammal & Associates, Inc.

Stewart, H.C., Jr. (1966), *Ground-water Resources of Polk County. (Florida),* Florida Geol. Survey R.I. 44.

Stringfield, V.T. and H.E. LeGrand (1966), *Hydrology of Limestone Terraces in the Coastal Plain of the Southeastern United States,* GSA Special Paper 93.

Thornbury, W.D. (1954), *Principles of Geomorphology,* John Wiley and Sons, New York.

Vernon, R.O. (1942), *Geology of Holmes and Washington Counties, Florida,* Florida Geol. Survey Bull. 21.

Vernon, R.O. (1951), *Geology of Citrus and Levy Counties. Florida,* Florida Geol. Survey Bull. 33.

Windom, S.R. and Campbell, K.M. (1981), "Sinkholes Follow the Pattern," Geotimes, August Issue, pp. 20-21.

Winston, G.O. (1975), "Florida's Ocala Uplift is not an Uplift," *Am. Assn. Petroleum Geol. Bull.,* 61, pp. 992

Wyrick, G.G. (1960), *The Ground-Water Resources of Volusia County. Florida,* Florida Geol. Survey R.I. 22.

SUMMARY OF CHAPTER 5

Two tectonic processes have actively contributed to the face of Florida today. Folding and fracturing are associated with plate tectonics and isostasy or regional loading stresses. The Ocala uplift, Chattahoochee anticline, and Sanford high are responsible for exposure of Eocene limestones within the state and exert a significant influence on modern day karst and sinkhole topography.

Florida has abundant groundwater resources. Four major aquifers are found within the state, the major one of which is artesian and underlies most of the state. The groundwater characteristics contribute significantly to soil subsidence and sinkhole formation processes. Marshland subsidence is associated with the drying of muck soils. Sinkholes in the northern section of the state basically result from cavern roof collapse, while in central Florida failure normally results from ravelling of the sands and clays overlying the limestone cavern system.

The Winter Park sinkhole resulted from a variety of geologic features, including intersecting fractures, interior drainage, lowered groundwater levels, and high limestone solution rates.

Conservation and protection of Florida's groundwater resources are of major importance. Pollution of our groundwaters routinely results from salt water intrusion and contamination from landfill disposal of industrial and urban wastes.

Name ______________________ Section ____________ Date____________________

REVIEW OF CHAPTER 5

Part A

1. The major tectonic processes molding the Florida peninsula ______________ are and ______________________________.
2. Large scale deformation is associated with ____________ processes while regional deformation results primarily from ______________________.
3. Which of the following is not a Florida anticlinal structure?
   a) Peninsular Arch
   b) Ocala uplift
   c) Chattahoochee Anticline
   d) Sanford High
   e) all of the above are anticlines
   f) all of the above are synclines
4. Which area of Florida contains no known crustal fractures?
   a) Florida Keys
   b) Everglades
   c) the area along the Peninsular Divide
   d) Panhandle
   e) all areas of the state contains fractures
   f) None of the areas above contain fractures
5. Normal faults are usually associated with
   a) tension
   b) compression
   c) lateral forces
   d) graben formation
6. Florida's major aquiclude is
   a) Peninsular Arch
   b) Ocala Uplift
   c) Hawthorne Formation
   d) Ocala Limestone
7. If the piezornetric pressure of an aquifer increases
   a) the water level of an artesian well falls
   b) the water level of an artesian well rises
   c) the solution rate of limestones decreases
   d) the acidity of the local wells increases
   e) both a and c are true
   f) both b and d are true

8. The Floridan Aquifer
   a) lies at a uniform depth beneath the entire state
   b) plunges to greater depths beneath the southern portion of the state
   c) becomes more acidic in coastal areas
   d) provides the major source of drinking water for central and south Florida
9. Shallow aquifers
   a) are usually non-artesian
   b) display a rapid response to heavy rains
   c) tend to reach a stable level during a drought
   d) are usually confined to a upper porous layer
   e) both b and c are true
   f) all of the above are true
10. Recharge areas
    a) tend to lie above areas of fractured limestone
    b) contain at least one spring
    c) normally lie above a thick and undisturbed aquiclude
    d) contain waters heavily basic due to the presence of cypress trees
11. Which of the following is indicative of a sinkhole prone area?
    a) presence of saltwater in the aquifer
    b) near surface piezometric level
    c) thick porous overburden
    d) extreme seasonal fluctuation of water table
12. Which of the following is not an important contributor to the state's saltwater intrusion problem?
    a) permeable limestone aquifers
    b) extensive coastal development
    c) population growth
    d) periodic droughts
    e) all are important factors
13. A common method of replenishing ground water levels disturbed by drought is
    a) control dams and canals
    b) increased ground water pumping
    c) drilling into the aquifer to depths greater than ordinarily required
    d) increasing lateral intrusion rates

Part B

1. How may faults be recognized without drilling?

2. What types of faults are recognized in the state?

3. Why are small faults often difficult to recognize?

4. A folded and faulted anticline has a flank dip angle of 45 degrees. Is significant subsidence expected in such a structure? Why?

5. What factors influence the hydraulic gradient?

6. What does a high chloride content in the aquifer indicate?

7. What features influence the patterns of cavern zones?

8. What type of sinkhole formation is most common in central Florida? Describe the process.

9. What important factors contributed to the formation of the Winter Park Sinkhole?

10. How are Florida's groundwaters becoming polluted?

11. What causes natural wells or solution pipes?

12. How is Miami's rapid growth affecting water availability from the Biscayne Aquifer?

CHAPTER 6

The Coastline—
Florida's Ring of Sand and Water

INTRODUCTION

Florida is surrounded on three sides by a ring of sand and water. This ring's boundaries shift continuously with the actions of the winds, the waves, and the currents. The zone where the land and waters meet is known as the **shoreline.**

The waves which roll along Florida's shores are the children of the wind. The size and power of the waves increases proportionally with the energy of the wind and with the distance over which it blows. Strong winds produce short, powerful waves with little regularity of pattern. These waves are called **sea waves**. Waves which have escaped the force of the wind and have returned to a more regular, cyclic pattern are referred to as **swells**. Differences between the characteristics of sea waves and swells are reflected in Figure 6-1.

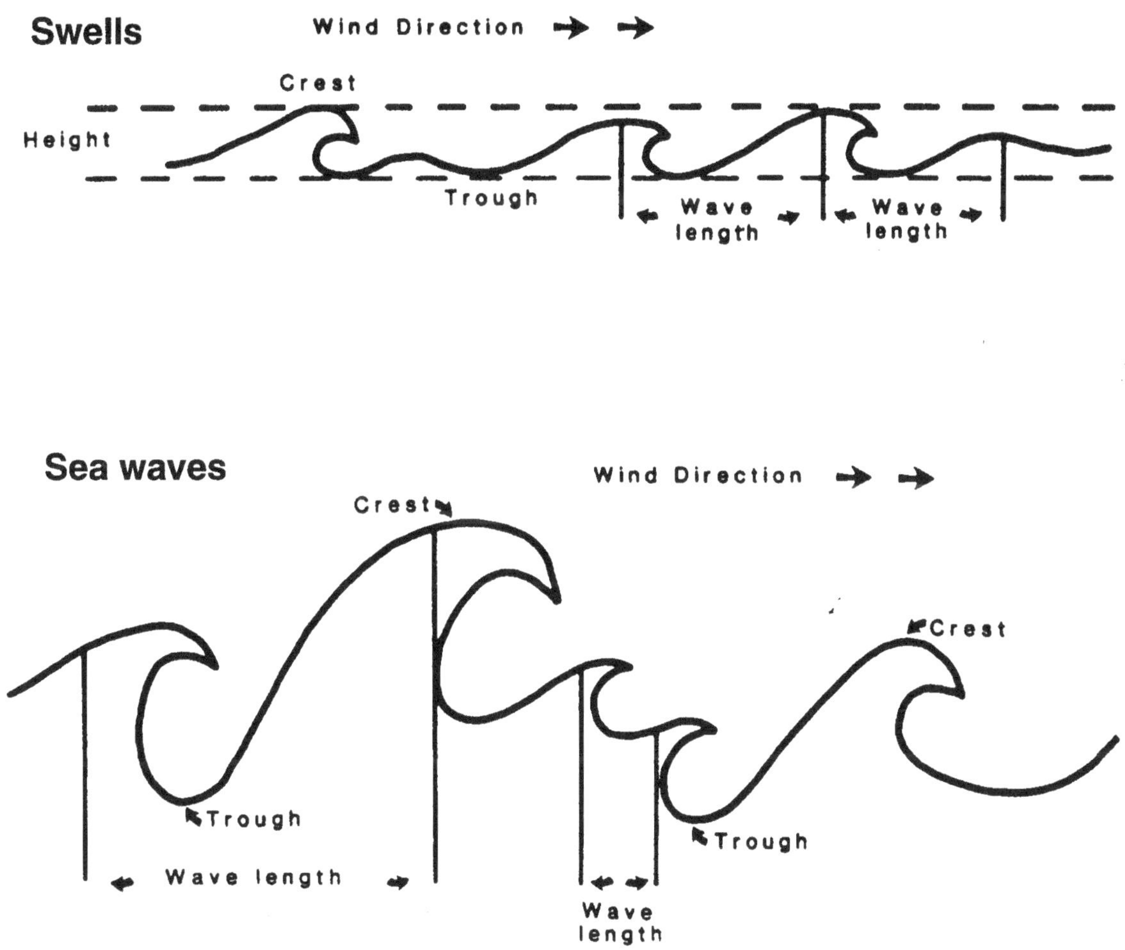

Figure 6-1 Comparison of wave patterns from swells and sea waves.

Sea waves notoriously crash onto the shore and erode much of what stands in their path. But it is the gentle swells that continually lap upon the shore which truly mold the Florida coast. The swells have given the state a coastline with an overall tidal exposure of approximately 9,000 miles. The Atlantic coast consists of long sand or sand and shell beaches stretching from the Georgia border to the tip of the peninsula. A series of barrier islands lie along the shore and protect the lagoons, estuaries, and river mouths from the damaging effects of sea waves.

South of Florida Bay, the warm waters promote the growth of coral and plant algae. The waves have mixed their shifting sands with cemented remains of this growth to create the Keys which stretch some 200 miles southward.

Tropical estuaries, salt marshes, and muddy coastlines extend westward from Florida Bay to Cape Romano. This is the area of the mangrove swamp formed along the drainage network of the Everglades. Here, dense foliage blocks out the Florida sun, and one must stand upon mangrove prop roots to escape the dark mud. From Cape Romano northward to Tarpon Springs, Florida's Sun Coast resembles the sandy beaches and barrier islands of Atlantic shores. Mangroves and mudflats again dominate the west coast from Tarpon Springs north to Cedar Key. The absence of beaches in this area may be attributed to two factors: (1) the bottom profile is very flat and shallow so that little sand is scoured from the bottom and carried landward, and (2) major streams like the Suwannee and Withlacoochee Rivers are spring fed and carry only negligible quantities of sand for deposit along the shore.

Westward to Pensacola the shoreline returns to a beach and barrier island character. Here, beach dunes and barrier islands are especially strong features, as the Apalachicola River provides sands far in excess of what the waves and currents can transport.

The topography of the continental margin beneath the waves is as transient as that of the coastline. The submarine landscapes reflect not only sand movement and deposition but also such chemical solution and accretion as we have associated with our limestone platform.

FLORIDA'S SHORELINES AND SEA COASTS

Florida's sea coasts encompass the zone along the water's edge. This is a zone of continuous change; no single feature is truly permanent. In a previous chapter we examined earlier coastal processes which culminated in Florida's interior sand ridges and dunes. However, to understand the changing topography of our coastlines we must delve more deeply into the mechanisms which have created Florida's beaches, keys, and coastal marshes as they exist today.

Beaches—Florida's Border of Sand

Wading in the shallow waves, one quickly recognizes the sands are in motion. It is these wave-swept sands which nourish and replenish our beaches. Thus, one familiar with the Florida shoreline would include this water zone within the boundaries or limits of the beach. In this context then, the seaward boundary of the beach extends to a depth of some 30 feet, beyond which wave energy is usually insufficient to transport significant amounts of sand.

Florida beach sands are derived naturally from five major sources. These include: (1) Appalachian and coastal lowland sediments carried southward by long shore currents; (2) ocean current erosion of the sea floor deposits; (3) mechanical breakup and disintegration of marine shells; (4) erosion of off-shore coral and algae reefs, particularly those destroyed during Pleistocene low sea levels; and (5) Pleistocene sand deposits and accretions disturbed by wave and current action along the Florida continental shelf. A minor source of sand has been attributed to offshore winds which blow loose sands from Florida's ridges and dunes. These sands may either be deposited along the shore or fall into waves which return them to the beach. Wind-blown sands are becoming more important as man's activities disturb the beach dunes and dry beaches. On a state-wide scale, Florida's rivers are also considered a minor source of sand, as their local rate of sand deposition is small when compared to the major processes described.

The materials which comprise Florida's beaches may be divided into three categories. **Terrigenous** sediments are granular particles eroded from the land and either deposited or carried in suspension in the sea. **Biogeneous** materials represent the shells of marine organisms, while **authigenic** materials are deposited on the sea floor as chemical precipitates.

Terrigenous sands and occasional heavy minerals tend to produce broad hard beaches with gentle slopes. The gradient or slope angle tends to increase along beaches composed of larger particle sizes. Along Florida's east coast, the average range of sand particle sizes is from 0.20 mm to 0.55 mm, while on the west coast the average sand particle sizes range up to approximately 0.70 mm. The effect of sand grain sizes is also recognized in other beach characteristics. In the case of Daytona Beach, the medium sized sands (0.21 mm average) provide the pavement-like surface used for auto racing. Larger grained sands of the Gulf Coast produce a softer and more pliable beach.

Biogenic beaches contain a mixture of sand and shell fragments in varying proportions. Abundant shell fragments tend to increase the average grain size of beach materials and to promote a steeper gradient. Along Miami Beach, the average particle size is 0.58 mm, and the beach gradient is significantly steeper than that of the more sandy beaches to the north. It is not uncommon along Florida's east coast to see abrupt changes in beach character from sands to shell litter. Sections of Cocoa Beach, for example, are covered with a "pavement of shells," while only a few feet away the sands are essentially void of even shell fragments. The reasons for this feature are unclear.

Along the coast of the Panhandle the beaches consist of white quartz sands with only small amounts of other materials. Shell fragments which promote the brownish color of most other beaches are absent. The lack of shells along Panhandle beaches is attributed to the steeper offshore slope and to extensive offshore bars which prevent landward movement of the shells by wave action.

The Profile of Florida Beaches

The profile of a beach (Figure 6-2) is controlled primarily by activity within the surf or littoral zone. Sand is washed up the beach face by advancing water (called **swash**) and is pulled back toward the sea by the returning water of **backwash**. The energy distribution of incoming waves determines whether the beach progrades (enlarges) under the advance of the swash or retrogrades (erodes) under the retreat of the backwash. Florida beaches

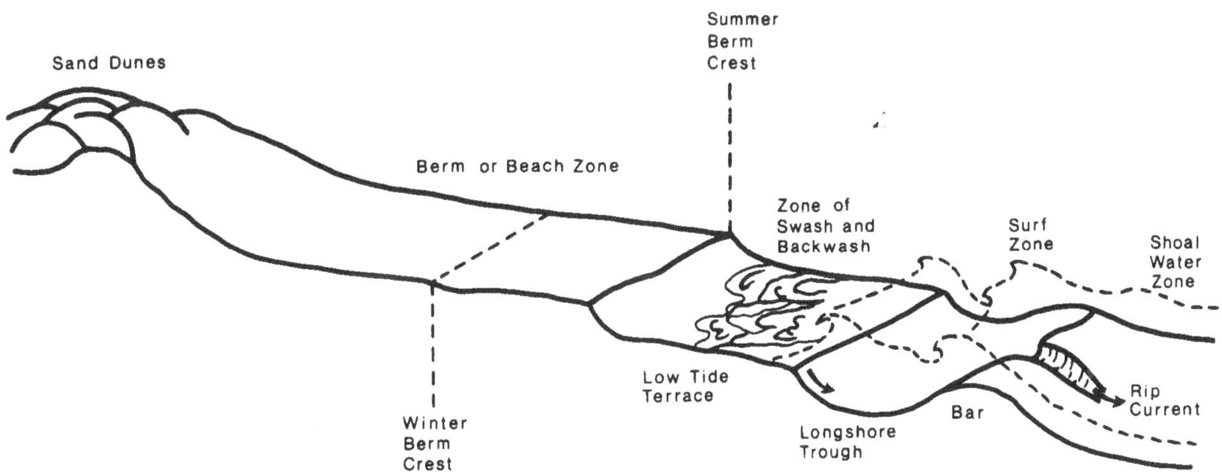

Figure 6-2. Profile of a typical Florida beach.

exhibit a seasonal tendency to prograde during the summer and retrograde during the winter months. During the Florida winter, local storms create close inshore waves with greater steepness and energy. Higher energy levels improve the wave's ability for sand suspension. Under these conditions, sands removed from the beach by backwash are not redeposited along the shore but are carried beyond the surf zone. Conversely, summer storms are more prevalent further out at sea. The resulting offshore storm waves disturb deeper sea floor sediments, stirring the sands into suspension, and carrying them landward. These waves weaken with distance, and, collapsing along the shore, they deposit their suspended sands. The net result is to repair the damage caused by the local winter storms.

The beach profile tends to vary with the season. Winter beaches are generally narrow and consist of relatively coarse sands (the finer sands are winnowed away). The steep high energy waves of winter usually erode a steep beach gradient, while the weaker waves of summer create a more gentle profile. Florida's summer beaches are typically 10 to 40 feet seaward of winter shorelines, and are characterized by finer sediments. However, sudden summer storms near shore may create waves which erode a scarp or step along the foreshore. Ultimately, however, the berm returns to its previous slope. A significant feature of Florida's beach profile is the presence of one or more offshore submerged ridges during the winter storm season. These submerged ridges are termed **offshore bars**. Offshore bars are the product of sea wave and higher swell erosion of the shoreline. The forceful backwash accompanying these waves deposits its sediment border where the water energy is lowered. Offshore bars can reach heights of several feet and extend for 10 miles or more in length. Investigators have found that beaches with profile gradients of less than one foot rise per 50 feet often exhibit multiple bars.

Once the offshore bars are established, the currents created by the backwash tend to flow somewhat parallel to the bar. However, the return of the backwash occasionally becomes so energetic that it is capable of breaching the bar at irregular intervals. These breaches then permit the current to flow out to sea. Currents around such breaches become particularly strong and are referred to as **rip currents**. The breach is termed a **rip channel**. Locally, rip currents may be referred to as undertow, although there is no significant downward pull to rip currents.

Longshore Currents

Just outside the surf zone the shallowing sea bottom drags upon the incoming waves and bends or refracts their path to one paralleling the shore. However, in most cases the refraction is incomplete. Thus, incoming waves strike the shore obliquely and are reflected down the beach; the result is a current running parallel to the shore. This current is referred to as **longshore or littoral drift**.

Sands suspended in the waves by surf zone turbulence and backwash are moved down the beach by the littoral drift. The path of movement is erratic due to the effect of individual waves. The movement down the beach is continuous, however, and results in a river of moving sand.

Along Florida's east coast the net littoral drift follows a southerly course. The speed of the drift is greatest during the fall and winter months when 30 to 50 mph winds routinely blow from the northeast. The Atlantic storms and winds promote large-scale erosion and sediment transport. Conversely, Florida's Gulf coast is more protected and the average sediment transport is perhaps one-tenth that along the Atlantic shores. Movement of Atlantic sediments between Fernandina and New Smyrna Beaches were found to be 400-500 thousand cubic yards per year. Comparatively, 50 to 100,000 cubic yards are moved annually along the lower west coast and 100 to 200,000 cubic yards along the west coast and Panhandle.

Florida's Cuspate Shorelines

Longshore currents and incident wave patterns have created a series of ridges along the beach line. The ridges form between the berm and low water mark. Many of these ridges appear transverse or inclined toward the shore, while others lie parallel to the beach. The ridges may extend thousands of feet in length and reach heights of up to six feet. Ridged shorelines projecting outward are termed **cusps**.

Cuspate-forming sediments are deposited when the energy of the waves or currents is interrupted. Such energy interruption occurs naturally (1) where slack water forms at the collision of current eddies; (2) at the mouth of bays where longshore current velocity slows and wave patterns change; (3) where long shore current velocity slows on the inward side of meanders or bends; (4) where wave and/or current energy is insufficient to transport excessive sediment loads provided by rivers and streams; and (5) where offshore shoals and organic reefs concentrate wave patterns. The formation of a typical cuspate feature is shown in Figure 6-3. Other common coastal features produced by current deposition of sediments are shown in Figures 6-4 and 6-5.

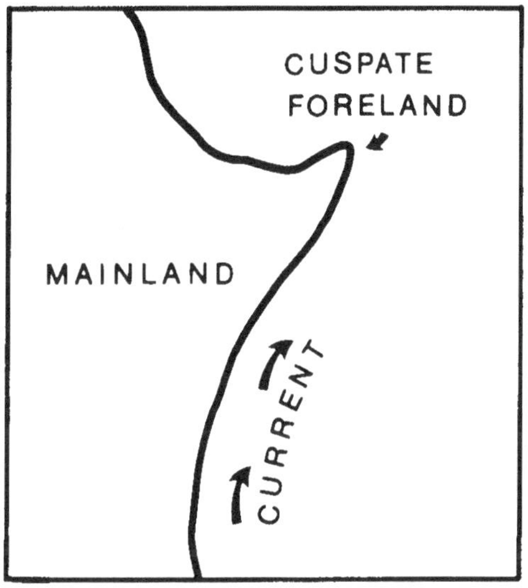

Figure 6-3. Formation of a typical cuspate foreland.

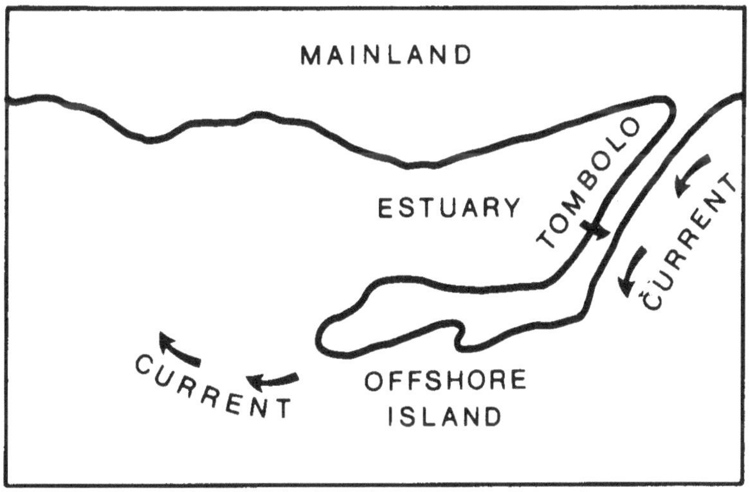

Figure 6-4. Tombolos, are the result of longshore drift depositing sands in such a manner as to create a connection between the beach and the offshore island.

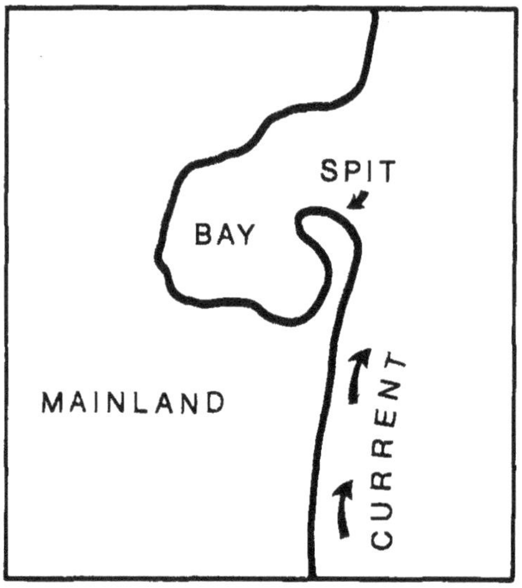

Figure 6-5. Spits result when sands are deposited near the mouth of a bay or harbor.

Specialized terms are applied to several of the cuspate forms. Large cusps more than a mile across are referred to as **cuspate forelands**. The terminology applied to smaller cusp-like features is generally characterized by the special process or feature which caused their formation. Florida's largest cuspate foreland is Cape Canaveral, which thrusts out into the Atlantic 14 miles from the mainland. The beach ridge angles southeastward from the mainland and is separated from it by the Indian and Banana Rivers. To the north and south, the foreland fades into the barrier islands flanking the peninsula. Cape Canaveral has been created by the northward flowing Gulf Stream as it swings eastward and away from the coast. Conversely, the northern flanks of the Cape experience significant erosion from northeasterly winter storms and summer trade winds.

Less known forelands occur along the Gulf Coast as well. Near the mouth of the Apalachicola River, Cape George and Cape San Blas have resulted from the combination of littoral drift and excessive sediments supplied by the Apalachicola River.

Retrogradation of Florida's Beaches

Many beach areas of the state are experiencing retrogradation or erosion due primarily to: (1) the gradual rise in sea level; (2) meandering of the littoral or longshore current; and (3) man's over-development of these areas. During the period from 1964 to 1972, this rise was measured at the average rate of one-quarter-inch per year. This corresponds to an annual loss of beach width of six to 30 inches. This rate of retrogradation is continuing unabated today.

Shoreline erosion is most pronounced where meanders of the littoral current swing landward. Here, the current energies are directed forcefully into the beach, where they remove the sands and transport them down the shore. Many of man's efforts to control the damages from littoral currents have actually promoted the erosion they were designed to control.

A case in point is the effect of groins and jetties. Groins are walls or embankments constructed at right angles to the shoreline with the intention of trapping the littoral sands on the updrift side. While this method is successful, little sand is captured on the leeward side. As the littoral currents swing around the end of the groin, they angularly attack the shore, producing additional downdrift erosion (Figure 6-6). The cure for this additional erosion becomes a second groin, and thus the process is repeated and prolonged until the beach is laced with groins. Jetties are similarly constructed for the purpose of protecting bays and inlets. Their effect on sediment accumulation and "downdrift" erosion is similar to that created by groins. The scope of such coastal construction is indicated by the fact that 14 of Florida's 21 Atlantic inlets are equipped with jetties and groins.

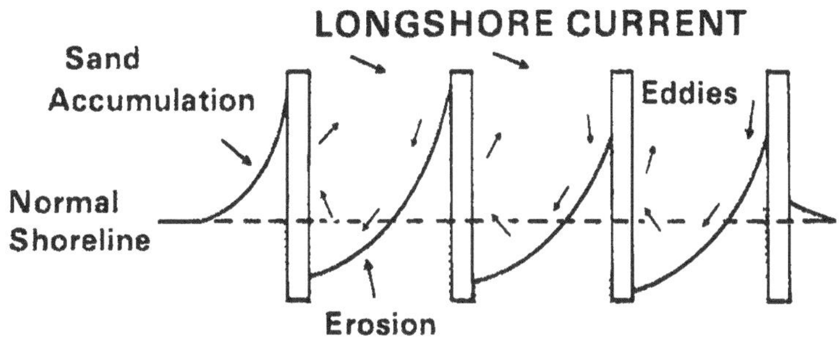

Figure 6-6. The effect of sand accumulation along groins and jetties is readily apparent. Sand collects on the updrift side of the groins, but tends to erode on the downdrift portion due to eddy currents.

Breakwaters are constructed parallel to the shore for the purpose of breaking up strong swells and sea waves before they assault the beach. They also provide the protection of an artificial harbor for small craft. In cases where the breakwater is physically connected to the shore, littoral drift flows around it to produce a spit on the downdrift side. Where no physical connection exists, littoral drift commonly deposits sands in the quiet water between the beach and the breakwater (see Figure 6-7). The result is that the shore grows seaward until reaching the protective breakwater.

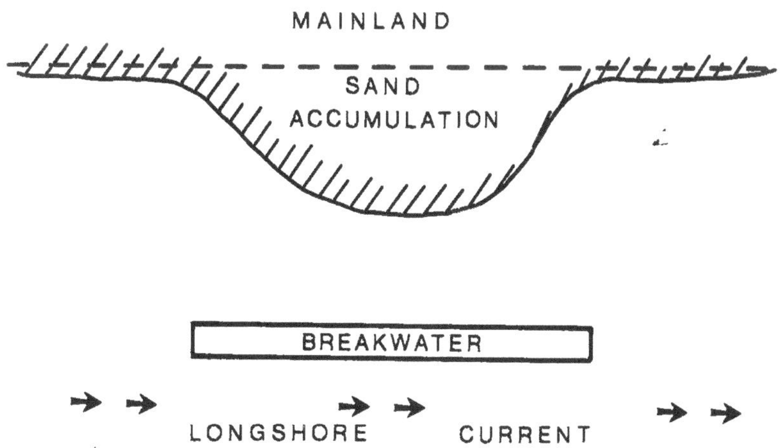

Figure 6-7. Sand accumulation behind a breakwater.

Groins, jetties, and breakwaters are constructed along Florida's northern Atlantic coast to protect the area from the annual winter attack of north Atlantic winds (northeasters). The area of northeaster attack extends from Jacksonville to Cape Canaveral. South of this cape, the Florida beaches are sheltered from the northeaster winds by the Bahama Banks lying well to seaward. South Florida, however, lies along a common path of tropical hurricanes and routinely experiences storm damage along the coast every three years.

Miami Beach is a classic example of man's misunderstanding of coastal erosion processes. Unfortunately, it is an example being readily followed in other areas of the state. For years, sand dunes along the shore provided a natural barrier to storm surge encroachment. Bulldozing of the dunes and their protective covering of vegetation has replaced nature's barrier with man's development. Bulldozed sands lack the stability of natural deposits and are subject to accelerated erosion.

Sea walls are routinely constructed to protect the hotels, condominiums, and other developments from the encroaching waves. Two common types are: (1) precast concrete slabs or sheet steel attached to embedded footings and land anchors (Figure 6-8), and (2) free-standing gravity structures with enlarged concrete bases. Both types of structures commonly fail from undermining or erosion by waves breaking over their tops.

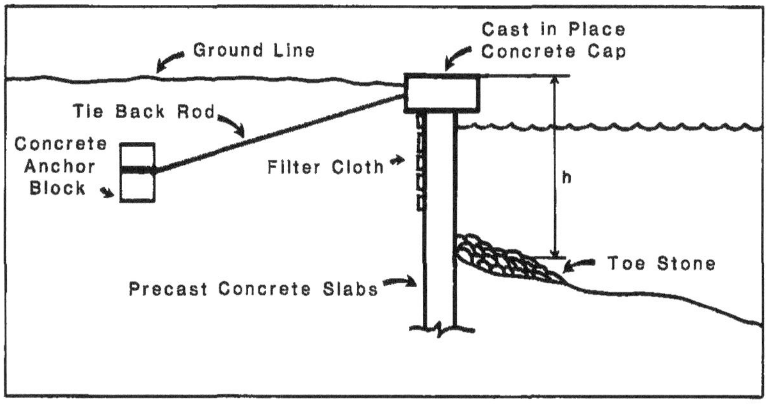

Figure 6-8. Features of seawall construction with embedded footings and land anchors. (Reprinted by permission of Florida Sea Grant Program.)

A study prepared by the U.S. Army Corps of Engineers indicates that over one quarter of Florida's sandy beaches are critically eroded. Along the Atlantic coast from Jupiter Inlet south to Miami's Baker Beach (a distance of over 100 miles) the shoreline regression is nearly 20 feet annually. Similar erosion rates occur along the Gulf coast from Lido Key northward to Clearwater.

Several erosion control methods are now employed which more accurately approximate nature's beach processes. These include:

1. **Artificial renourishment** in which sands are pumped from offshore barrow pits (up to a mile in length and 15 feet deep) piled into mounds along the shore. The mounds are then flattened by earth-moving equipment to create the desired beach widths and profiles. The renourished berms reach widths of up to 250 feet and increased heights of up to 10 feet. Artificially renourished beaches do not have the life of natural ones primarily due to compaction and sea bottom imbalance differences.

Some renourished beaches last only a few days (as in the case of Manatee County's Anna Maria Island Beach); others have lasted up to 10 years before renourishment was again required.

2. **Vegetative plantings** reduce erosion from surface water runoff and swash encroachment. Sea oats (Figure 6-9) and tall Bermuda grass work well along beaches with steepness ratios of less than 3:1.

Figure 6-9. Sea oats along a west coast beach (Reprinted by permission of Florida Sea Grant Program.)

3. Construction of **stone revetments** (Figure 6-10) parallel to the shoreline presents a strong and rough surface which is capable of reducing the impact energy of breaking sea waves along the beach. To be most effective, the revetments are often underlain with filter cloth and fine gravel.
4. Construction of **Makepeace wood groins** extending offshore permit the waves to dissipate their energy as they flow through holes in the groin. Such dissipation of wave energy causes the waves to deposit their suspended sands as they gently roll up onto the beach. The result is an actual increase in beach width on either side of the groin. Makepeace wood groins have been in use in Florida for over 30 years without failure of either groin or protected beach.

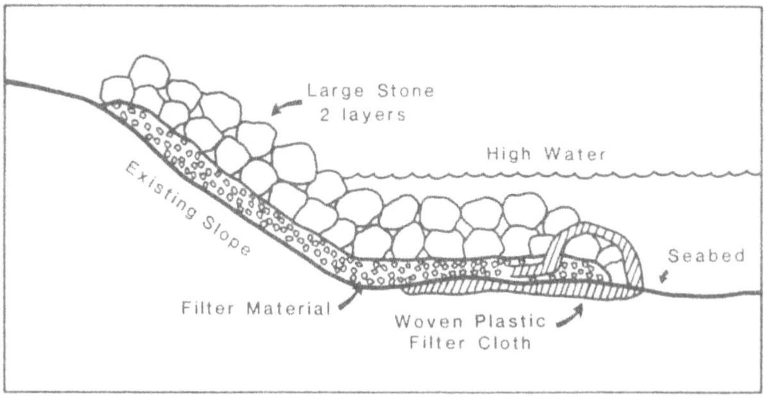

Figure 6-10. (Top) Schematic placement of stone revetments over filter material for shoreline protection. (Bottom) Concrete slab revetments protect the seawall on the left from damage by smashing waves.

Additional measures to control beach erosion include building code restrictions, zoning regulations halting improper stabilization processes (i.e., jetties and breakwaters), movement of threatened buildings, and stricter enforcement of other pertinent environmental practices. Public education of the effects of unregulated growth and development is helping to overcome many of the deficiencies described. Typical features experienced by man-made seawalls are shown in Figure 6-11.

Figure 6-11. Two types of seawall failure. (Top) Failure resulted from breaking off of land anchor connections. (Bottom) Erosion by waves breaking over the top undermined the wall. (Reprinted by permission of Florida Sea Grant Program.)

FLORIDA'S MANGROVE COAST

Florida's Gulf Coast is the home of two extensive mangrove forests. The forests are basically salt water swamps whose trees are anchored in the shifting muds by a tangled mass of roots. Man often regards mangrove forests as a place of mystery. He sees them as a place of water stained dark by tannin, a place with a jumbled jungle of intertwined branches, or a place of shadowy and clogged waterways winding between ten thousand islands.

Yet Florida's mangrove coasts serve several important purposes. The mangrove roots trap and hold the sediments of Florida's mud flats and offshore bars. The roots also catch and hold the myriad shell fragments, driftwood, and aquatic debris continually moved about by currents and waves. Shorelines thus become stabilized and protected from the erosive forces of waves and storms. Progressively, a solid-like shoreline evolves, although the soils are a diverse mixture of materials, and contain high salt and acid sulfate levels.

The mangrove root zone is the home of a wide variety of marine creatures. Uncounted millions of tiny crustaceans, oysters, mussels and worms feed upon the litter filtering to the bottom from the overhanging branches. In turn, these creatures become food for a multitude of bottom feeding fish, crabs, and wading birds. A similar host of feeders thrive among the mangrove leaves and branches.

Mangrove coasts are also important to Florida's fishing and seafood industries. These coasts are the breeding ground and preserve for shrimp and a variety of fish. In the Tampa Bay area nearly half of the mangrove coastline has been converted to subdivision housing. The result has been a significant decrease in both fish and shrimp production throughout the entire bay area.

Three species of mangrove inhabit the Florida coast. The seedlings of each type are moved about by the moods of the waves until they find desirable lodging. **Red** mangroves are capable of colonizing sand bars, mud flats, or any debris accumulations existing in normal sea water. The stilt-like prop roots of red mangroves arch out from the trunk and branches. They multiply rapidly to provide an anchor in the waves. Seedlings sprout within the roots of their parents and gradually expand the island of stability. The mangrove islands enlarge, link with one another, and ultimately provide a unified front to the waves. Red mangroves growing along the Keys are generally shrub-like, but along Shark River they reach heights of over 80 feet.

Preferring shelter from the storms and waves, black mangroves take root landward of their red cousins. **Black** mangroves prefer brackish, less saline waters and lack stilt-like prop roots. Their rooting system more nearly resembles that of fresh water cypress trees complete with breathing knobs above the water level. In general, black mangroves are short scrub trees, but occasionally they achieve heights overshadowing their red mangrove neighbors.

Small **white** mangroves root inland from their fellow species. They prefer a relatively diluted sea water The landward edge of the mangrove forest is marked by a gradual appearance of cedars, buttonwoods, and shallow-rooted palms.

If left undisturbed, mangrove forests may be expected to enlarge the Florida mainland. Hurricanes, tornadoes, and tropical storms, however, routinely uproot the advancing edge of the mangrove coasts and destroy entire forest sections. The forests also promote their own destruction through the flocks of birds they attract. Guano dropped by the birds is fatal to the tissue of the mangroves, and ultimately contributes to the death of the trees. Thus, rookeries of ibis, spoonbills, egrets, and pelicans eventually become a forest of skeleton trees, devoid of foliage and life.

The State of Florida and the Army Corps of Engineers are now working diligently to protect the mangrove coastline. Under Florida law all mangroves below the mean high water line are considered the property of the state, while the U.S. Army corps of Engineers administers the forests to landward. The two organizations are now actively controlling man's development and commercialization of this coastal resource.

FLORIDA'S CONTINENTAL SHELF

Florida is surrounded by a shallow, terraced platform called the **continental shelf.** The platform extends gently seaward until it terminates in a somewhat steep, plunging embankment called the **continental slope.** In 1946 a presidential decree defined the edge of the continental slope as beginning at the 100 fathom mark, when in fact the slope's beginnings are irregular.

Significant differences exist between Florida's continental shelf along the Atlantic and Gulf coasts. Along the Atlantic seaboard, the continental shelf off the city of Jacksonville is about 60 miles wide and some 50 fathoms deep. As the shelf extends southward it becomes progressively more narrow and shallow. Off Palm Beach the shelf reaches a width of about one mile and shoal depths of 15 fathoms. Florida's continental shelf is the home of numerous creatures such as the manatees of Figure 6-12.

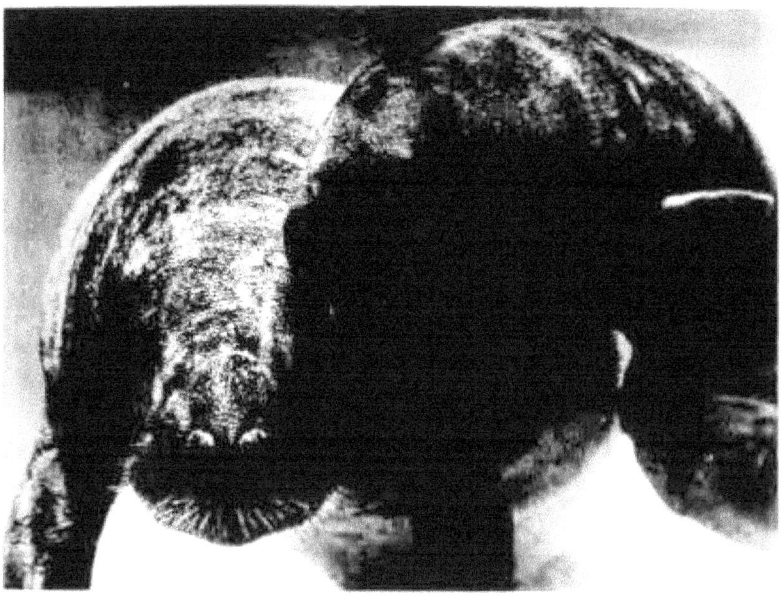

Figure 6-12. Manatees at play along Florida's Gulf Coast. (Reprinted by permission of Florida Game and Fresh Water Fish Commission.

The Atlantic Shelf slopes off onto a broad flat Blake Plateau, which stands as a margin or shelf overlooking the deep Bahama Basin. The Blake Plateau is separated from Florida's Atlantic Shelf by a 125 fathom trench. Along the western edge of this trench the Blake Plateau supports coral banks reaching 150 feet high. However, most of the Plateau seaward is a flat, somewhat featureless plain. The trench is attributed to the erosive action of the Gulf Stream and its prevention of sediment accumulation.

The Gulf Stream hugs the coastline south of Miami and has virtually removed the shelf across the southern end of Florida. The Keys stand upon narrow coral reefs which overlook a trench of 100 to 500 fathoms. In Florida Bay between the Keys and the mainland, the bay floor is generally less than 10 feet deep. The bottom consists of the mud deposits and luxurious growths of marine grasses attached to the limestone floor. It is these muds which support the mangrove coastlines previously discussed. Beyond Florida Bay, however, the shelf width expands rapidly to about 100 miles. The shelf width remains relatively considerable. The western edge of the shelf is marked by an apparent series of faults which step sharply downward into the Gulf Basin. The floor of the basin is relatively flat and reaches depths of up to 2,000 fathoms (12,000 feet). Florida's Gulf coast lacks the erosive action of a strong current, and thus the shelf bottom is more gentle and more conducive to marine growth.

The character of Florida's continental shelf has been largely influenced by the manner of its formation. During the close of the last glacial epoch, an enlarged Florida peninsula was nearly encircled by a series of coral reefs. Along the Atlantic coast the reefs probably extended northward to near St. Augustine; in the Gulf the encirclement was complete. Erosion sediments filled the basins created by these coral dams, creating our Continental Shelf. The Shelf sediments are generally less than 15,000 years old. Sediments overflowing the coral dams described formed the continental slopes. The stability of the slopes is dictated in turn by the repose angle of their sediments, current activity, and bottom stirring that accompanies hurricanes.

The coral dams described are relatively recent by geologic standards and lie atop others dating from up to 130 million years ago. Each epoch since the Cretaceous has seen the creation of similar algal or coral dams and the subsequent sediment accumulation of continental shelves and slopes.

Along the Atlantic Shelf, sand ridges occur in fanlike patterns close to shore. Shoals (shallows) extend well across the shelf seaward of the major capes and cuspate forelands. Occasional hills and Pleistocene sinkholes appear some what randomly along the shelf.

The west Florida shelf is primarily a smooth plain marked by sporadic coral or algal reefs. The algal reefs are more prevalent along the outer portions of the shelf.

The sediments of the Florida shelf contain varying mixtures of sand and shell fragments, with pockets of calcareous oolites. The near shore sediments are primarily terrigenous sands, while authigenic or chemical deposited phosphorites are common along the shelf edge. Only sediments lying close in to shore and influenced by wave action exhibit grain size segregation (smaller sizes tend to collect seaward).

THE GULF STREAM AND THE FLORIDA CURRENT—OUR RIVERS IN THE SEA

Since man first encountered the Gulf Stream he has visualized it as a great "river in the sea." This concept helps one to understand some of its features, but the concept is not truly correct.

Off the Atlantic coast of Florida lies the Sargasso Sea. The Sargasso lies along a latitude where the sun stands high in the sky throughout the year. The Sargasso Sea thus annually receives a comparatively large amount of heat. For the earth to maintain its energy balance the heat energy received by the Sargasso Sea must be transferred to polar areas. This energy transfer process has resulted in a circular motion of the warm Sargasso waters relative to the colder adjacent seas. Such a circular, or swirling movement of water is called a **gyre**. A system of currents flows around the junction of the Sargasso gyre and colder water masses. This current system is called the **Gulf Stream.** The system consists of several current segments including the North Atlantic, Canary, North Equatorial, Antilles, Florida, and Loop currents.

The Gulf Stream system flows clockwise about the Sargasso Sea. This clockwise rotation is produced by friction with the prevailing surface winds, which also absorb much of the Sargasso heat. These same winds tend to push the currents toward the North American coast (Figure 6-13).

The volume of flow of the Florida current as it leaves the Gulf of Mexico is tremendous. Under the influence of powerful east winds the current is forced through the narrow channel between the Florida Keys and Cuba. The effect of this channel is to produce an increase in sea level of about 8 inches along the Florida coast and 18 inches along the Cuban coast. North of the Bahamas the Floridian current is joined by the Antilles current to form the Gulf Stream with a volume nearly twice that of the Floridian current. The maximum flow of the Gulf Stream along the Florida coast has been estimated to be nearly 5,000 tons per minute. This flow is more than 25 times the combined flow of all the world's rivers together.

The Gulf Stream continues north until reaching Cape Hatteras, where it leaves the continental shelf and swings into the North Atlantic. This seaward path frees the current from the confines of the coastline and allows the current to follow a more random course. The current begins to meander or swing into broad and irregular loops. Occasionally these loops break free of the main current and form rings of warm or cold water.

The rings rotate at the rate of two to three knots and drift away from the main Gulf Stream current. Cold-core rings 90 to 180 miles in diameter routinely drift south through Florida's Atlantic fishing areas. Cold-core rings are a boon to ocean fishing as they transport rich planktonic life which attracts large schools of fish. East of Florida these cold water eddies rejoin the main current.

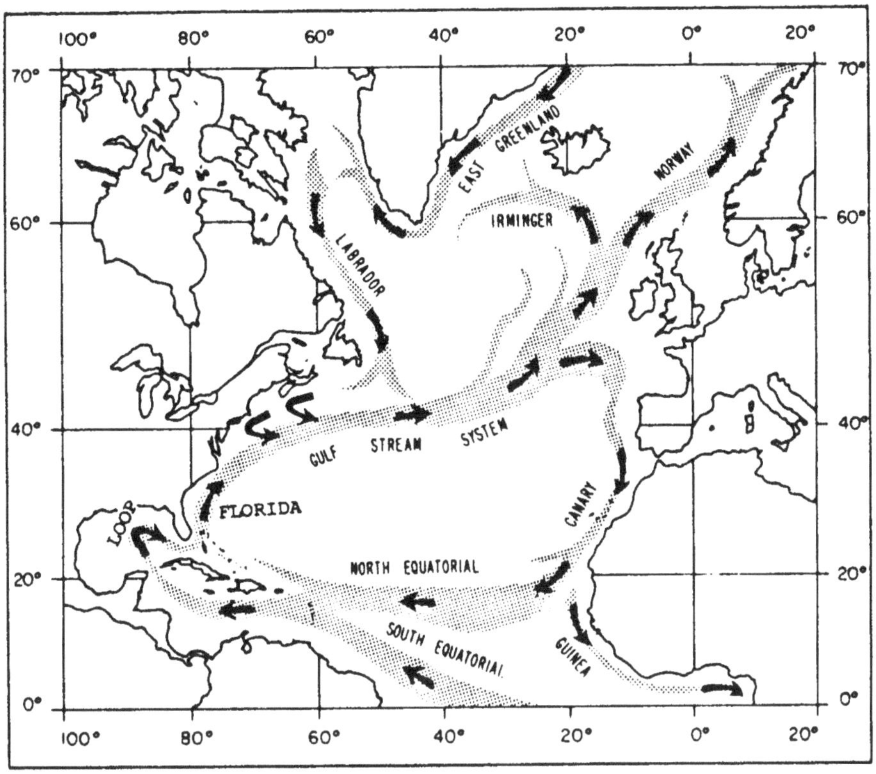

Figure 6-13. The Gulf Stream Gyre including its various currents. The path of this system passes along Florida's eastern and Gulf coasts. (From Oceanography by Jerome Williams. Reprinted by permission of Jerome Williams.)

Similar rings or eddies are formed in the Gulf of Mexico by the loop of current. Here recorded rings have measured from 60 to 240 miles in diameter. A variety of biological and chemical disturbances accompany ring flow onto Florida's shallow Gulf Shelf. Included in these disturbances are destructive plankton blooms and significant salinity and turbidity changes.

Water rings have recently been found to affect coastal weather patterns. Warm water rings encountering colder waters rise, creating the equivalent of a low pressure zone in the atmosphere. Cold winter winds sweeping over such an eddy are fueled with both thermal energy (as is a hurricane) and water vapor. A storm so created struck Florida's eastern seaboard Thanksgiving Day 1984. It devastated the eastern coast from Daytona to Miami and left a grounded freighter (ship) astride the swimming pool of a Palm Beach home. Cold water rings conversely produce the equivalent of atmospheric high pressure cells.

Warm water rings tend to swirl counter clockwise while cold rings rotate in an anticyclone (clockwise) manner.

Another little understood ocean phenomena is the waterspout. Waterspouts are weaker, shorter lived and travel over the water with slower speeds than do their tornado cousins. Both tornados and waterspouts occur in the right flank of squall lines. In a typical waterspout a funnel cloud swirls downward from a low cloud base (usually at heights of 2000-2500 feet above the waves). The funnel is made visible by atmospheric condensation from above. Within the funnel is an eye similar to that of a hurricane, though normally 200 feet in diameter or less. Where the funnel meets the ocean surface a tube of water rises about the funnel cloud.

This tube is a circular wall of sea spray raised by updrafts following the exterior of the funnel. Wind speeds about the vortex have been recorded at 170 mph and maximum funnel diameters near the cloud of about 150 feet. The typical funnel cloud moves over the waves at about 30 mph.

Waterspouts may be more common than generally believed. Most funnel clouds appearing over the ocean seldom reach the surface of the waves. Approximately one-third are accompanied by lightening and thunder. The average life of a waterspout is 15 minutes, although some have been observed to last for almost three hours.

Waterspouts are common along the northern Gulf coasts. They usually form during the summer months, and are most prevalent in Florida during June when the combination of warm ocean surface and sea breezes appears optimum.

RED TIDES ALONG THE GULF COAST

Red tide is a typical "scare word" for residents of Florida's Gulf Coast. **Red tide** represents the massive proliferation of certain marine algae in large, concentrated masses called "blooms." The most common algae associated with red tide are dinoflagellates, which occur in such numbers that the waters are discolored bright red to dark brown. Algal toxins associated with these blooms produce windrows of dead fishes killed by the poisons and washed ashore by advancing waves. A suffocating and irritating gas is often associated with these blooms.

Red tides have been particularly detrimental to west Florida's coastal based economy. During nine months from November 1946 through August 1947 a massive algae bloom covered approximately 150 miles of shoreline and reached some 40 miles into the Gulf. Several hundred million marine creatures were estimated to have washed ashore along the beaches. Business and public buildings near the shore were forced to close due to the sea breezes conveying noxious ocean spray inland. This was the worst red tide on record, though they have occurred regularly through Florida's history.

Actually red tides are common throughout coastal areas of the globe. Red tides generally occur in areas where strong, upwelling or storm induced turbulence carries algal-filled sediments into areas favorable for rapid plant division.

No economically feasible method of controlling red tide is currently available. There is even considerable debate as to whether red tide should be controlled. Several economically important marine species tend to multiply rapidly (e.g., shrimp) after a red tide incident, and become more abundant than before. There is some evidence that dinoflagellates may also be responsible for the formation of several recognized deposits of offshore oil.

REFERENCES

Anikouchine, W.A. & R.W. Sternberg (1973), *The World Ocean—An Introduction to Oceanography*, Prentice-Hall, Inc., Englewood Cliffs, NJ.

Bascom, W. (1960), "Beaches," *Ocean Science*, Readings from Scientific American, W.H. Freeman & Co., San Francisco, Calif.

Collier, CA (1975), *Seawall and Revetment Effectiveness, Cost and Construction*, Florida Sea Grant Program, Report #6, SUS of Florida, Tallahassee, FL.

Coleman, B.A., et al., (1986), "Red Tide: A Recurrent Marine Phenomenon," *Sea Frontiers, Sea Secrets*, Vol. 32, No. 3.

Eicholz, N. (1982), "Florida's Rivers, Priceless Gifts from Nature," *Florida Wildlife Magazine* (Nov-Dec 1982) pp. 31-35, Game & Fresh Water Fish Commission, Tallahassee, FL.

Emery, K.O. (1969), "The Continental Shelves," *Ocean Science*, Readings from Scientific American, W.H. Freeman & Co., San Francisco, Calif.

Shepard, F.P. (1963), *Submarine Geology*, Harper & Row, Publishing, NY.

Williams, J. (1962), *Oceanography*, Little, Brown & Co., Boston, MA.

SUMMARY OF CHAPTER 6

Florida is surrounded by a ring of sand and water with a coastline of approximately 9,000 miles. Florida's beach sands are derived from Appalachian sediments, sea floor deposits, shell fragmentation, erosion of offshore reefs, and disturbance of Pleistocene deposits. The qualities of the beaches vary with location and beach profile.

Beach profiles themselves vary seasonally and are generally steep and narrow during the period of attack by winter storms. Longshore currents, however, provide sand nourishment and contribute to the creation of a variety of beach forms such as cuspates, spits, and tombolos. Beach profiles experience significant modification due to man's activities such as groins, breakwaters, and seawalls. Man has, however, also learned to repair much of the beach retrograde action he has caused.

Florida's mangrove coast occurs along Florida Bay and from north of Tampa Bay to Cedar Key. The mangrove growth tends to stabilize the shoreline as well as promote marine life cycles.

The Gulf Stream system impacts both the state's continental shelf and beach activities. The Gulf Stream represents the flow about the Sargasso gyre. Meanders of the current often produce warm or cold water rings, which in turn affect life along the coast as rings drift by or collide with the shore.

Name ____________________ Section ____________ Date____________________

REVIEW OF CHAPTER 6

Part A

1. Florida's beaches and coasts are shaped primarily by
   a) the wind
   b) strong sea waves
   c) gentle sea waves
   d) off shore currents

2. Florida's mangrove coastlines occur
   a) in areas where extensive beach erosion is prevalent
   b) where streams deposit large amounts of sand adjacent to their mouths
   c) where bottom profiles are flat and stream deposition is negligible
   d) both a and b are true
   e) both b and c are true

3. Florida's shoreline is considered to extend into shallow depths (up to 30 feet) because
   a) this represents the low tide level
   b) this includes the zone of sand transport
   c) this includes the zone of stream sediment deposition
   d) this includes the zone of marine life
   e) both c and d are true

4. Along Florida's beaches
   a) sand accumulates continuously
   b) sand removal is decreasing beach sizes continuously
   c) sand accumulates during the winter and erodes during the summer
   d) sand accumulates during the summer and erodes during the winter

5. Rip currents tend to be strongest where
   a) back wash zones are small
   b) off shore bars lie close to shore
   c) off shore bars are breached
   d) surf zones are strong

6. Cuspate forelands tend to form
   a) where current eddies and longshore current interruptions are common
   b) at the mouth of bays where longshore currents are strong
   c) where wave patterns are weak around shoals and reefs
   d) where current velocity is strongest in meander bends
   e) both b and d
   f) both a and c

7. Beach erosion is common
   a) where longshore currents swing seaward
   b) where longshore currents swing landward
   c) on the updrift of groins and jetties
   d) inland of breakwaters
   e) all of the above are common erosion sites

8. South Florida beaches experience little erosion from northeastern winter winds because
   a) prevailing winter winds come from the Gulf
   b) the warmth of the Gulf stream reduces the effects of the cold northern winds
   c) the eroded sands are quickly replaced by Gulf Stream sediments
   d) They are protected from the storm by the Bahama banks

Part B

1. What are the sources of Florida's beach sands?

2. What three categories of sediments occur along Florida beaches and how do these categories differ?

3. What process creates and modifies Florida's offshore bars?

4. How do the longshore currents along the Gulf and Atlantic coasts differ?

5. What factors contribute to beach retrogradation?

6. How can retrogradation be controlled?

7. How does the character of the mangrove coast change over the years?

8. How have Florida's continental shelves been created?

9. What is the effect of warm and cold water rings upon the Florida coasts? What is the cause of these rings?

10. Which of the following is not considered a productive method of beach erosion control?
    a) Bulldozing of beach dunes
    b) beach renourishment
    c) makepeace groins
    d) revetments

11. Which of the following is an important accomplishment of mangrove forests?
    a) anchoring offshore bars
    b) stabilize shorelines
    c) act as a breeding ground for fish and shrimps
    d) retain mud flat sediments
    e) both b and a

12. The continental shelf along the Gulf coast is home of more marine growth than is the Atlantic shelf because
    a) erosive action of a strong current is lacking
    b) bottom stirring by hurricanes is more common
    c) the sand ridges along the Atlantic shelf inhibit coral growth
    d) the Atlantic shelf contains sinkholes while they are completely lacking in the Gulf.

13. The clockwise path of the Gulf stream is caused by
    a) coriolis, effect
    b) equatorial heating effects
    c) wind friction
    d) cold water rings

CHAPTER 7
Soil and the Soil Processes

INTRODUCTION

Soils are the most important of all resources associated with the earth's crust. In Florida, soils rank second only to climate as the state's most important natural resource. Without soil all other resources would be useless because there would be no human population—no one to utilize the resources.

All states and nations would prefer to be able to feed their people without depending upon outsiders for food. However, this is not possible because states and nations do not all possess fertile soils and suitable climates to produce the quantities and varieties of food needed to feed all of their population.

SOIL COMPOSITION

Soil can be defined as a mixture of mineral and organic substances which have developed over long periods of time by the weathering of rocks and the decomposition of vegetation.

The crust of the earth originally consisted of fire-born igneous rocks. After cooling, the rocks slowly broke into smaller pieces through a process known as **weathering**. Over longer periods of time, the smaller particles of rock further broke down into sediments. The sediments in turn became the source of a second group of rocks known as sedimentary rocks. As time passed, both the igneous rocks and the sedimentary rocks were subjected to great heat and pressure and were changed into a third group of rocks called metamorphic rocks. These three groups of rocks constitute the parent materials from which the mineral matter portion of soils is derived.

Igneous rocks comprise approximately 95 percent of the total rocks of the earth's regolith. In the solid state, igneous rocks are of little value for soil; however, upon weathering, the smaller particles become an important constituent of soil. One of the most abundant and widely distributed igneous rocks is granite. Granite consists largely of quartz and feldspar usually with mica and some other minor accessory minerals. Upon weathering, the quartz forms silica sand and the feldspar evolves into clay. Sand and clay are the two most important mineral constituents of soil, a third being limestone.

Sand comprises the largest particles found in soil and forms the framework for the more active soil components. Individual particles can vary in size from .05 mm to 2.0 mm. Because of the large size of the sand particles, the sand provides the largest areas of space between individual grains, allowing movement of air and water between particles and drainage of water within the soil.

Silt particles are smaller than sand particles and range in diameter from about .05 mm to .002 mm. They are able to fit into the pore space between sand particles. Since silt weathers faster than sand and retains water better, it releases more available nutrients and retains more water for plant growth than sand. Most silica sands are composed of silicon dioxide while silt consists predominantly of silicate minerals.

The smallest particles of mineral matter found in soil consist of clay, with diameters of the particles smaller than .00008 mm. Clay particles are composed of either silicates from weathering in the middle latitudes, or iron and aluminum hydrous-oxide clays in the tropical climates.

Limestone is geologically classified as a chemical sediment and is usually deposited by precipitation from water at the bottom of a shallow body of water. Thick layers of limestone can accumulate from prolonged evaporation of a restricted body of sea water.

Because soils consist of different mixtures of **mineral matter**, various names are given to certain combinations of the minerals. If the mineral matter comprising the soil sample consists of at least 85 percent sands and the remaining 15 percent clay and silt in a ratio of 1.5 to 1.0, the mixture is called a

sand. However, with a decrease in the amount of sand to less than 52 percent, the mixture is known as a **loam**. If the sand content is further decreased to less than 45 percent, the silt less than 40 percent and the clay increased to 40 percent or more, the soil is termed a **clay soil.**

THE ORGANIC COMPONENTS OF SOIL

Earlier, soil was defined as a mixture of mineral and organic substances developed over a long period of time by the weathering of rocks and the decomposition of vegetation. In this section the decay of vegetation will be discussed.

Although soil usually contains only about five percent or less of organic material, it constitutes a very important constituent of soil. Both fauna and flora inhabit the soil, varying in size from minute bacteria to vertebrate animals as large as prairie dogs and reptiles as large as the Florida gopher turtle. Bacteria are among the smallest inhabitants of the soil, yet are responsible for such important functions as decomposing plant and animal life and making nitrogen available to higher forms of plant life. Animal life, fauna, is also important as a component of soil. Earthworms are able to loosen soil, providing more pore space and making it easier for plants' roots to penetrate the soil. They also add organic waste to the soil.

Organic matter refers to all living and dead organisms making their home in or on the soil. Organic matter, called humus, consists of all types of plant litter in various stages of disintegration. However, it is usually considered as that part of the organic matter remaining after the major portion of the residue has decomposed. Humus increases the soil's ability to hold moisture, makes the soil easier to till, improves the soil structure, and provides an important source of carbon and nitrogen.

In addition to mineral matter and organic matter, soils also contain air and water, both of which occupy the pore space in the soil. The quantity of air in the soil at any given time varies inversely with the quantity of water in the soil. If the soil holds all of the water of which it is capable, it is saturated and it allows no room for air in the soil. Air consists mainly of oxygen, nitrogen and carbon dioxide. The oxygen is essential for living organisms which cause the decay of organic matter. Nitrogen is also necessary to support these organisms in addition to nourishing plants whose dead tissues will become humus. Carbon dioxide in the soil reacts with water, forming carbonic acid which aids in the dissolving of mineral matter.

Water must enter the soil before it can move downward and occupy the pore space of the soil. The process of entering the surface of the soil and moving downward is called **infiltration**. Upon entering the soil, the water moves downward by the force of gravity and forms a film of moisture upon each soil particle with which it comes in contract. Some of this water then becomes available to plants. Soil moisture is also very important as a soil-forming agent.

SOIL PROFILES

Soil is continuously undergoing change in its natural setting. Over a period of time, soil usually develops a series of layers called horizons. These layers vary in depth, structure, color, texture, moisture availability, mineral composition, and degree of acidity or alkalinity (Figure 7-1).

Lying beneath the A horizon is the B horizon, which begins with the zone of accumulation (illuviation). A common term, hardpan, is frequently applied to this zone as the zone tends to be impervious to the downward movement of water. The B horizon usually contains little organic material, but it is definitely weathered. Because the A horizon is usually less compacted and the soil structure more porous, water can percolate through the pore spaces, carrying with it finer particles of silt and clay. This process is called **eluviation.** At depths where the soil is more compacted and the distance between soil particles is too small for particles of silt and clay to move downward, the materials carried from the A horizon accumulate. The mineral matter that is dissolved in water or weak acids and carried to the zone of accumulation is due to the process of **leaching**. Because some minerals (limestone, iron and aluminum silicates) are more soluble in water and weak acids than quartz or silica sand, they are more susceptible to leaching. This can readily be observed in soil profiles by the, accumulation of the red iron compounds present in the illuviation zone.

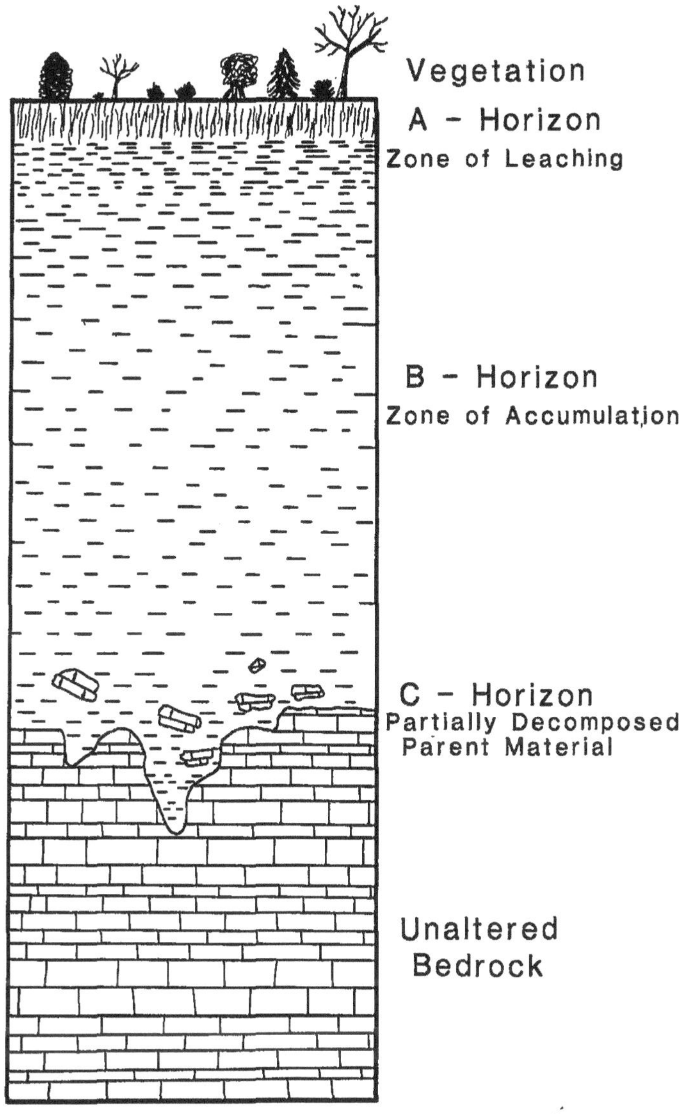

Figure 7-1. Diagram of a soil profile.

The A and B horizons are known collectively as the **solum**. Both contain weathered mineral matter. Beneath these horizons lies the C horizon which may contain relatively unadulterated minerals which may be either similar to or different from the minerals found in the solum. The soil processes that have contributed to the formation of the A and B horizons have little, if any, effect upon this horizon. Underlying the C horizon is consolidated bedrock designated by the letter R. A vertical, two-dimensional slice through a soil plug is called a **profile**. Such a plug is taken from the soil by an auger which is thrust into the ground, usually to a depth of 48 inches. The circular core is removed from the soil, sliced longitudinally and placed into a semicircular container. The horizons can be seen along the longitudinal slice.

Some soils have evolved *in situ* (in place) from the parent bedrock underlying horizons A, B, and C. Such soils are referred to as residual soils, as contrasted with soils which have been transported to the present location by various gradational agents, called transported soils.

FLORIDA SOILS

The entire state of Florida lies within the landform region of the United States known as the Coastal Plain. The Coastal Plain is characterized by almost flat terrain, sedimentary parent materials, and variable soil drainage. The seemingly flat terrain, which is typical of the land along both the Atlantic and the Gulf coasts, is actually a series of terraces, varying in elevation from sea level to approximately 100 feet. The drainage systems are usually not well defined, and their effects upon soil formation have been important factors in developing the properties of the existing soil.

With the exception of the areas adjacent to the major streams where alluvium has been deposited and in regions covered with organic deposits, the soils of Florida have been derived from marine-deposited, fine, sandy clays overlying limestones at varying depths.

Florida soils have been developed in, and exposed to, a climate consisting of mild winters, hot, humid summers, an average annual rainfall exceeding 50 inches, and a high percentage of possible sunshine. The factors of climate and vegetation have helped produce highly leached soils with little organic material on the surface. Thick layers of humus, except in the Everglades and similar areas, have not accumulated because of the rapid oxidation brought about by warm, humid climate and by animal life and organisms living in and on the soil.

SOIL CLASSIFICATION

The previous section discussed the general development of Florida soils. In this section the manner in which soils are categorized and classified will be discussed. Classification is a method of taking vast quantities of information, categorizing the information into a valid number of generalizations, and presenting these generalizations in a logical and practical manner.

Soils have been classified in some manner for hundreds of years, and there is some evidence that the Chinese formulated a simple soil classification based upon structure and color as early as 4,000 years ago. Most early classifications were based on broad soil groupings, such as agronomic use (how specific crops reacted to certain soils), organic matter content, color, texture, and moisture content (Caldwell et al., 1974). Each of these classifications served a specific need; however, they lacked the scientific reliability that could be measured quantitatively by one pedologist (one who studies the soil) and repeated by another with the same or similar results.

One of the earliest of the modern attempts to scientifically classify soils began in Russia about 1870 under the auspices of a Russian by the name of V.V. Dokuchaiev. Dokuchaiev viewed soil development as a dynamic system which involved the interaction of parent material, plant and animal life, geomorphic factors, and time.

It was not until the beginning of World War I, in 1914, that European and American pedologists became familiar with the soil research being carried out by Dokuchaiev and his students. One of the Americans who became familiar with these studies was C.F. Marbut, Chief of the United States Soil Survey. Over the following quarter of a century Marbut developed a genetic soil classification that culminated with his 1938 classification in which he used eight great soil groups and adopted many Russian terms for his groupings.

During World War II, the United States supplied most of the food to feed its Allies. The increased demand for food showed the need for a greater knowledge of soils, and as a result, the United States Government sponsored many soil research projects. Before World War II, too much emphasis had been placed upon soil under virgin conditions, and as a result failed to adequately include soils that had been eroded, such as the soils of the Dust Bowl of the 1930s. The Marbut classification did not appear broad enough to encompass such soils.

In 1951, a new classification system was begun. This system proposed to provide a workable method classifying soils anywhere in the world. Under this proposed system, soils were grouped together according to their physical, chemical and mineralogical characteristics. Between 1951 and the date the system was adopted by the United States, January 1, 1965, the new classification underwent seven major revisions or approximations.

In the United States Comprehensive Soil Classification, soils are classified under six broad levels of generalizations based upon differences in measurable and visual characteristics of soil horizons. The presence or absence of these characteristics indicates the dominant active soil-forming process or the lack of such development.

The first level, called the Order Level, has the fewest number of similarities present as compared to the lowest level, Series Level, where there is a high degree of homogeneity in both soil features and soil genesis. The four levels, between the Order and Series levels are: Suborder, Great Group, Subgroup and Family. The degree of homogeneity increases with each lower level.

Many of the 10 soil Orders fall into definite geographic ranges, indicating the importance of climate in this classification of soils. Of the 10 groupings under the Order Level, nine are mineral soils while one consists of organic matter. Statewide, Categories I (Order Level), V (Family Level) and VI (Series Level) are more important to Florida farmers and growers than Levels II, III and IV. Category V (Family Level) identifies soil properties or characteristics which are important to soil use and management, such as plant root growth, textural characteristics, mineralogy, thickness of horizons, and soil temperature and reaction. Soil temperature refers to the average soil temperatures at depths of 20 inches below the surface.

In Florida two soil temperature zones characterize the soils, Thermic and Hyperthermic The boundary between these two temperature zones can be seen on the Soil Map of Florida, Figure 7-2. North of this boundary the term Thermic Temperature Zone is applied. The Thermic Temperature Zone includes soils with a mean annual temperature of between 59°F (15°C) and 72°F (22°C) at a depth of 20 inches below the surface. South of this boundary lies the Hyperthermic Temperature Zone with its soils having a mean annual temperature of more than 72°F (22°C) at a depth of 20 inches below the surface. The soils of both the Thermic and Hyperthermic Temperature Zones have temperatures that vary more than 9°F (5°C) between mean winter and mean summer temperatures.

Category VI, Series Level, is the lowest level of the six categories, and the soils in this group exhibit the most complete homogeneity in profile characteristics, even though the texture of the surface soils may vary from place to place. The strength of the United States Comprehensive Soil Classification System is based upon differentiating properties that soil scientists can observe in the field to distinguish one series from another or to be able to infer characteristics that differentiate between series. With the Marbut Classification of 1938, much emphasis was placed upon soils under virgin conditions. Under this system, emphasis is placed on soil depth beneath the land where the soil has been plowed.

SOIL ORDER NOMENCLATURE

Specific names are applied to soils in each of the six category levels. However, in this chapter, names will be used only for the 10 Order levels of Table 7-1 and of these 10 Order levels, only the four Order levels pertaining to Florida will be discussed and shown on the Soil Map of Florida (Figure 7-2).

The mineral soils in a particular Order level occupy that particular category largely because of the degree of weathering that the soil has undergone. Some soils, like Entisols have no horizons, or only the beginning of one, while the Ultisols are the most weathered soils of the mid-latitude.

The U.S. Comprehensive Soil Classification System has many advantages over the older classifications. The characteristics by which the soils are classified are ones that can be observed in the field by one soil scientist, verified by a second, and measured quantitatively. The subjectiveness of the older classifications have been replaced by more objective criteria. A second advantage lies in the nomenclature. The names of the Order levels are related to the soil characteristics of the particular Soil Order. For example, the word "histos" comes from the Greek, meaning "tissue." The decayed vegetation comprising the Histosols has its origin in the once-living tissue of plants that had grown in place but later died and decayed.

Ultisols

Ultisols are among the most weathered soils of the humid sections of the mid-latitudes, generally low in fertility because of the shortage of bases and characterized by a subsurface clay horizon which has accumulated by the action of percolating soil water and the long-term influences of vegetation and climate.

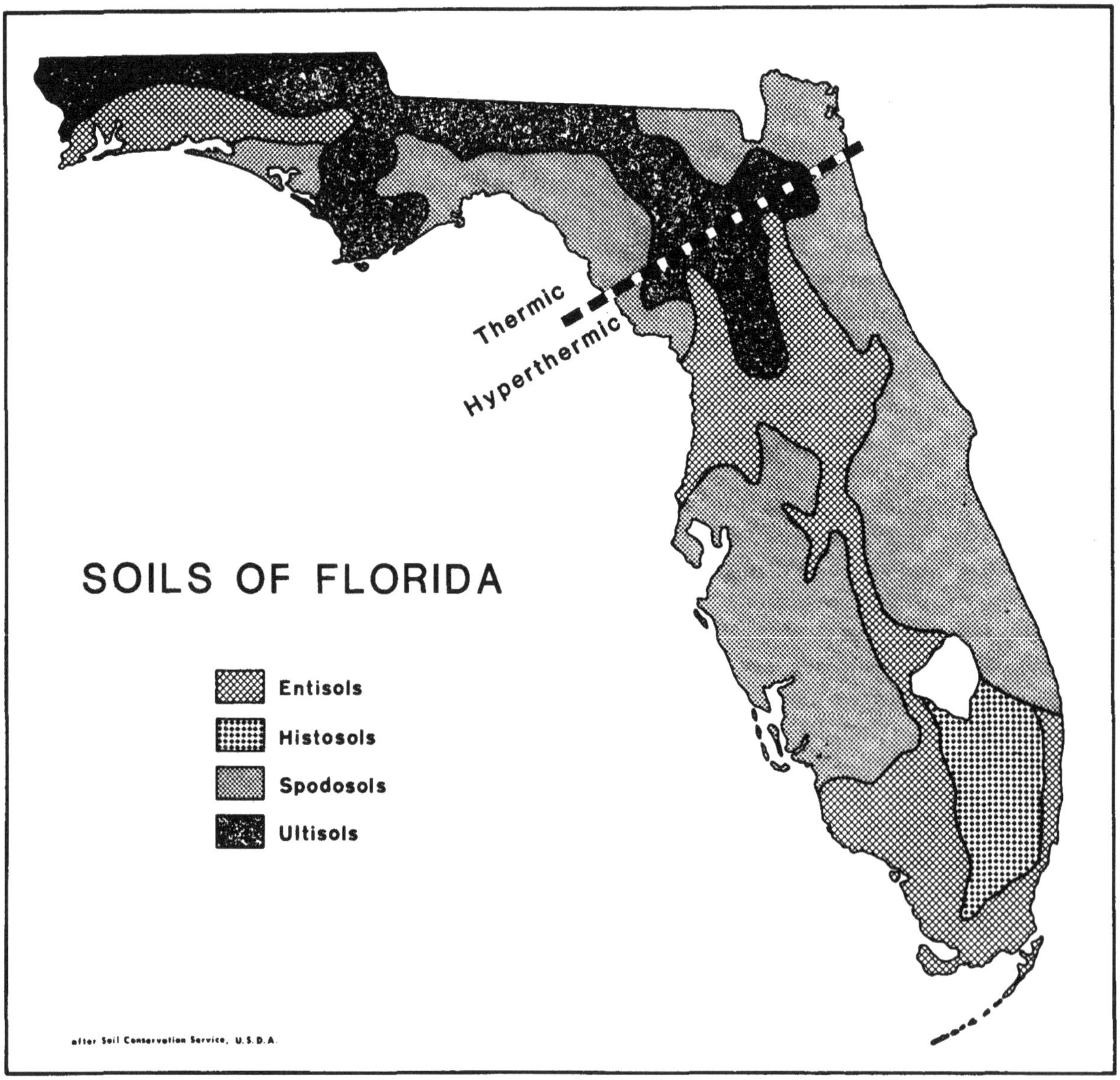

Figure 7-2. Soil map of Florida. (USDA.)

The Ultisols occupy an area that coincides closely with the northern highlands as far south as Marion County. Except for the flood plain of the Apalachicola River, they do not reach into the coastal lowlands. For the most part, the Ultosols are gray, grayish-yellow, reddish-brown or brownish-gray at the surface. In the subsoils, which are heavier in texture than the surface soils, the usual colors are red, yellow, or red and yellow mottled. They have developed principally upon the marine deposits of noncalcareous sands and clays, with some developing on limestones. Because the topography of this area is rolling, it is more subject to erosion than any other part of Florida.

The sandy clays possess good natural fertility which can be utilized for general farming and some special crops. The farm products grown in these regions are pasture grasses, some corn, oats, and wheat, soybeans and tobacco. The forests consist mainly of pine trees that are used for both lumber and for pulp, from which paper is made. Hardwoods, such as oak and hickory, are also grown on these soils and harvested for flooring and lumber for furniture.

Table 7-1. Derivations of Soils Order Nomenclature

Soil Order	Language Root From Which Word Derived	Meaning of Formative Term	Percent of Florida Soils	Percent of World Soils
Entisol	———	*Ent* from "recent"—a nonsense syllable	24.5	12.5
Vertisol	Latin	*Verto* meaning "turn"	0.0	2.1
Inceptisol	Latin	*Inceptum*, meaning "turn"	0.00	15.8
Aridisol	Latin	*Aridus* meaning "arid" or "dry"	0.0	19.2
Mulltisol	Latin	*Mollis* meaning "soft"	0.0	9.0
Spodosol	Greek	*Spodos* meaning "wood ash"	37.7	5.4
Alfisol	———	*Alfi* from "pedalfer" Al = aluminum Fe - iron	0.0	14.7
Utisol	Latin	*Ultimus* meaning "last" or "ultimate"	31.4	8.5
Oxisol	French	*Oxide* meaning "oxide"	0.0	9.2
Histosol	Greek	*Histos* meaning "tissue"	6.4	0.8

National Cooperative Soil Survey, *Soil Taxonomy* (1970), U.S. Government Printing Office, Washington, D.C., pp. 3-17.

Spodosols

Spodosols are mineral soils that must contain a Spodic or Placic horizon overlying a fragipan subsurface of clay and organic matter. The Greek word "spodos" means "wood ash," which describes the ashen-gray color of the surface soil of the Spodosols. The ashen-gray color minerals consist of aluminum and silica, with some organic matter, but little or no iron. They accumulate during the process of eluviation as the soluble minerals are leached out and the suspended minerals are eleuviated to a lower horizon. (Eluviated horizon is one that had soil material removed in suspension from the soil overlying it.)

The Spodic horizon referred to in the definition to Spodosols lies below the ashen-gray minerals of the upper soil and consists of the accumulation of clay with smaller quantities of aluminum and carbon. The Placic horizon differs from the Spodic horizon in that the accumulated material appears to be cemented together by iron which gives it a color of red to black.

Underlying either or both the Spodic and Placic horizons is the fragipan, commonly called hardpan in Florida. The prefix "fragi-" is derived from the Latin term "fragilis," which has the English connotation of brittle. The word pan applies to a layer or horizon in the soil of clay or similar materials. Thus, a fragipan is a brittle pan consisting of an illuvial horizon lying beneath the soil surface, consisting of materials that have been precipitated from around water solution or deposited from suspension.

Spodosols are found in both cold and warm climates. The largest expanse of this Order Level is found in the subarctic boreal forests of North America and Eurasia. The boreal forest is a large homogeneous expanse of needleleaf conifers that can withstand the extremely long cold winters and low evapotranspiration of the subarctic climate. Spodosols found in Florida have evolved in a warm, humid climate; however, the dominant type of vegetation is the needleleaf evergreen conifers, predominantly pine in the drier sections (Figure 7-3), and bald cypress, a needleleaf non-evergreen conifer, in the wetter areas.

Figure 7-3. Planted pines in flatwoods showing Spodosol soil under the trees.

The poorly drained flatwood soils of Florida contain the largest expanse of Spodosols in the state. They are located along either coast and characterized by gray to black surface materials with the parent materials varying from limestone to sand or clay. The topography is usually level or depressed with a high water table and low natural fertility. Along the east coast the soil is predominantly sterile coastal sand. The natural vegetation that grows in this area gets much of its nourishment from the rain and ground water. Pine trees and either scrub oaks in the north or saw palmetto in the south form the natural vegetation pattern of the region. Spodosols can be managed for agricultural use with the proper application of lime to decrease the soil's acidity, the addition of trace elements and deep plowing and tilling. Agriculturally, the soils are used mainly for grazing animals and growing vegetables, potatoes, strawberries, and tropical fruits.

Entisols

Entisols are mineral soils that lack definite soil horizons and profile. This may result from the immaturity of the soils or from some aspect of the physical environment that has prevented the development of horizons. Newly deposited beach sands or recently deposited alluvium readily fit into the Entisol Order.

Entisols are found on every continent in the world because they can exist in practically any moisture region on any type of parent material and under almost any form of vegetation. In Florida they constitute almost 25 percent of the state's soils, dominating the soils of the dry, almost sterile sands of the ridges of the Central Highlands and the coastal areas of southeast and southwest Florida (Figure 7-4).

Figure 7-4. Deep, sandy Entisol soil in the Central Highlands near Citrus Tower.

In the Central Highlands, the soil consists of well to excessively drained sands which may extend to depths of many feet. The percentage of silt and clay is very low in this region. However, citrus trees will thrive when water and the proper nutrients are supplied. Moderately to excessively drained sandy soils are also found along the coastal lowlands of the Gulf Coast from the Alabama line to the Apalachicola River. These soils developed on beds of non-calcareous sands over beds of acid sandy clays. Such soils are not inherently fertile, but they make good land for watermelons when irrigated and fertilized.

Along the southwest coast, in the vicinity of Ten Thousand Islands and Cape Sable, the Entisols consist of poorly drained soils, with cypress the dominant type of natural vegetation. The soils vary widely in color depending upon the underlying materials and the region where the soils are located.

Almost all of Florida's soils are generally acidic; they are not very fertile; they lack humus; and much of the time they are either too wet or too dry. Man, however, has learned to work with the soils by adding lime, fertilizer, and trace elements (minerals). To meet these problems man has to make adjustments in the way that he farms in Florida. In the dry, sandy Entisols, the farmer has turned the disadvantages into advantages. He has done this by draining the sandy fields, using chemicals to kill the animal and vegetative life in the soil, and finally placing the plant in the ground to be fed and watered as necessary to produce a profitable crop. The soil merely holds the plant in place.

Since the Entisols are light, loosely packed soils, Florida farmers have used technology through the development of plants that have a root structure especially adapted to Florida conditions.

Histosols

The Histosols constitute the only non-mineral soil Order. Both worldwide and in Florida, the percentage of the total acreage classified as Histosols is the smallest of all of the 10 soil Orders. Some 6.25 percent of Florida's 35 million acres are in Histosols, placing Florida fifth among the states in land area covered by peats and mucks. Roughly three-fourths of the state's organic soils are in the Everglades (Figure 7-5), with smaller areas scattered throughout the state. The largest of these outliers is near Zellwood, adjacent to Lake Apopka, just north of Orlando (Figure 7-6).

Figure 7-5. Histosol peats and mucks near Belle Glade.

Figure 7-6. Histosols near Lake Apopka in Zellwood.

The Greek word "histos" means "tissue," the main component of organic matter. By definition, Histosols contain organic matter varying from 12-18 percent organic carbon by weight and exceeding 50 percent by volume. They can develop in any climatic regime as long as water is available and in their natural setting they are saturated. Only when they are drained and during droughts do they become dry. When this happens they compact and oxidize.

Peats and mucks have similar origins and differ mainly in the degree of composition of the vegetable matter. The plant material is largely undecomposed and can be recognized in peat soils, while the decay is so far advanced in the muck soils that the origin cannot be identified. Because of the large quantities of decayed vegetation, these soils may look very black and rich but they are frequently highly sandy and need additional trace minerals before they can be effectively utilized. When drained and managed correctly, Histosols are important to agriculture. In the Lake Okeechobee region, valuable vegetation is grass (actually sedge, not grass) and cypress trees. Special soil islands in the Everglades are common. These include bare limestone outcrops (Figure 7-7) and islands of dry, more fertile land which grow palms, pine trees and, in some cases, oaks and other hardwoods. In the river valleys where the land is flooded only part of the time, these soils support oaks, cypress, hickory and gum trees. Several major problems arise from the drainage of the Histosols. One is the problem of oxidation which occurs when the organic matter is exposed to air. The oxygen in the air actually causes the organic matter to oxidize, and the soil is lost. For this reason, it is not uncommon for a farmer to either leave a heavy cover crop, such as clover, on the field or to flood the field when it is not in use. Another problem is that of burning. When the soil is dry, the organic matter catches fire rather easily and often burns for long periods of time.

Figure 7-7. Bare limestone at surface of land near Mahogany Hammock in Everglades National Park.

REFERENCES

Baldwin, M., C.E. Kellog and J. Thorp, (1938), "Soil Classification," *1938 Yearbook of Agriculture: Soils and Man*, U.S. Government Printing Office, Washington, D.C.

Batten, J. and J.S. Gibson, (1977), *Soils*, University of Alabama Press, University, Alabama, 276 pp.

Bryan, O.C., (1962), *Soils of Florida and Their Crop Adaptation*, Bulletin No. 42, State Department of Agriculture, Tallahassee, Florida, 48 pp.

Buckman, H.O. and N.C. Brady, (1971), *The Nature and Property of Soils*, McMillan Company, New York.

Buol, S.W., F.D. Hole and R.J. McCracken, (1973), *Soil Genesis and Classification*, The Iowa State University Press, Iowa, 376 pp.

Carlisle, V.W., C.E. Beeman, J.H. Herbert, Jr., and E.L. Hinton, (1974), *Soil Identification Handbook*, Soils Department, University of Florida, Gainesville, Florida, 242 pp.

Gaddum, L.W. and H.L. Knowles, (1953), *Our Physical Environment*, Houghton-Mifflin Company, Boston, Massachusetts, 623 pp.

Hunt, C.B., (1972), *Geology of Soils: Their Evolution, Classification and Uses*, W.H. Freeman Company, San Francisco, California.

Kellog, C.E., (1936), *Development and Significance of the Great Soil Groups of the United States*, U.S. Department of Agriculture Miscellaneous Publication No. 229, United States Printing Office, Washington, D.C.

——— (1963), "Why a New System of Soil Classification?" *Soil Science*, Vol. LXXXXCI, No. 1, pp. 1-5.

National Cooperative Soil Survey, *Soil Taxonomy* (1970), U.S. Government Printing Office, pp. 3-17.

Steila, D., (1976), *The Geography of Soils*, Prentice-Hall, Inc. Englewood Cliffs, New Jersey, 222 pp.

Thornton, G.D., W.W. McCall, R.E. Caldwell, and F.B. Smith, (1960), *Soils and Fertilizers*, Bulletin No 137, State Department of Agriculture, Tallahassee, Florida, 79 pp.

Trudgill, S.T., (1977), *Soil and Vegetation Systems*, Clarendon Press, Oxford, England.

SUMMARY OF CHAPTER 7

Soil can be defined as a mixture of minerals and organic substances which have developed from the weathering of rocks and the decomposition of vegetation. Sand comprises the framework of soil and provides the largest areas of space between grains. The smaller particles of silt and clay occupy the space between the particles of sand.

Over periods of time, most mature soils develop a series of layers called horizons which vary in depth, structure, texture, color, moisture availability, mineral composition, and degree of alkalinity or acidity. The horizons, above the parent bedrock, are A, B, and C. A vertical two-dimensional slice through these horizons is known as a profile.

Florida soils have developed on an almost flat terrain of sedimentary parent material and variable soil drainage. With the exception of alluvium along major streams, the soils were deprived mainly from marine-deposited, fine, sandy clay overlying varying depths of limestone.

On January 1, 1985, the United States Soil Survey adopted the United States Comprehensive Soil Classification. Under this classification soils are classified into six Category Levels of generalizations based upon differences in measurable and visual characteristics of soil horizons. The six levels are numbered I-VI and named: Order, Suborder, Great Group, Subgroup, Family, and Series. There are ten groupings within the Order level, nine of which are mineral soils and one organic matter. Of the ten groupings only four are found in Florida.

Name ________________________ Section ______________ Date______________________

REVIEW OF CHAPTER 7

Part A

1. Water infiltrates more rapidly in sands than in clays because
   a) the sand particles are smaller
   b) clays and silts consist predominantly of silicate minerals
   c) clays contain iron and aluminum hydrous oxides
   d) sands contain larger particles and larger pore spaces
   e) both clays and silts retard infiltration

2. Match the following by placing the corresponding letter in the blank on the left.

____	1.	mixture of organic and mineral substance	A. clay
____	2.	disintegrating organic material	B. humus
____	3.	water movement through soil	C. coastal plain
____	4.	smallest particles in a soil	D. infiltration
____	5.	chemical sediment	E. soil
____	6.	area of sedimentary parent materials	F. limestone
____		and flat terrain	G. loam
____	7	soils formed in place from weathered rock	H. in situ.

Part B—Map Exercise

In the appendix there is a blank soil map to be completed. Using the following legend, complete your map.

Soil Order	Color
Entisol	Yellow
Histosol	Black
Spodosol	White
Ultisol	Red

Part C

1. What are the dominant factors that are responsible for the development of Florida soils?

2. Differentiate between sand, clay, and loam soils.

3. What areas of the state have soils with the largest quantities of humus in the soil? What areas of the state are devoid of humus. Why are they devoid?

4. Define the following terms:
   a) Eleuviation

   b) Illuviation

   c) leaching

   d) zone of accumulation

5. What is the difference between a soil horizon and a soil profile? Give an illustration of each.

6. Difference between thermic and hyperthermic temperature zones. What influences do soil temperatures have upon soil?

7. What area of Florida includes soils with the best developed profiles? Why are they located in this region?

8. What type of soil is found underlying cypress swamps? What happens to this soil when it dries?

CHAPTER 8

Florida's Tropical and Subtropical Vegetation

INTRODUCTION

When the Europeans first came to Florida, trees blanketed almost the entire northern part of the state from the Perdido River to the Atlantic Ocean. On the peninsula, lakes, ponds, and grasslands provided interruptions to the almost endless cover of trees. Yellow pine and cypress dominated the species of trees with fewer stands of hardwoods.

Cypress thrived in the wet areas, and islands of hardwoods occupied the hammocks (Figure 8-1). In the wet hammocks along the low energy coast, from Apalachicola to Tampa, the red cedar grew in abundance.

Figure 8-1. The Senator, a cypress tree, dated at between 3,000 and 3,600 years old near Longwood, in Seminole County. This is the oldest known tree in the State of Florida.

In general, trees will grow in any area where water is available throughout the year. If there is a definite winter dry season of at least two or three months, trees will give way to grasses. Climate thus becomes a factor that influences the type of natural vegetation. Soil also has its influences upon the type of native vegetation. Hardwood trees grow better on the heavier soils that contain clay and hold moisture. Most of the hardwoods in the middle latitude regions drop their leaves during part of the year and go through a dormant stage. The leaves are usually broad, and the wood is relatively hard and heavy. Some broadleaf trees do not lose their leaves during part of the year. The magnolia is a hardwood tree that has both broad leaves and is an evergreen. Some species of oaks, such as the live oak and water oak, are hardwoods that lose their leaves and develop new ones within a very short period of time.

Needleleaf trees, usually evergreen and softwood, can grow on soils and under conditions that most hardwood trees cannot tolerate. The narrow needleleaf does not give off as much water by transpiration as does the broader leaf of the hardwood tree. As a result, needleleaf trees can survive with less precipitation and on drier soils than the hardwoods. They will grow on clay soils, but eventually they are squeezed out by the hardwoods because they are shielded from the sun, which they need to grow, and the climax vegetation becomes predominantly hardwoods.

GENERAL CHARACTERISTICS OF FLORIDA VEGETATION

Small areas, or lens, of heavier soils scattered throughout regions of lighter sandy soils, support islands of hardwood trees known as **hammocks**. In these hammocks oak, maple, hickory and magnolia dominate while the adjacent areas support various species of pine trees.

Frequently trees do not reach full growth because of lack of sunlight, deficiency of water or infertile soil. Such stunted trees are referred to as **scrub**. On the deep sandy ridges of the Central Highlands, scrub oak trees have replaced the pine that were cut by lumbermen during the past half century. One of the largest areas of scrub trees in the state is in eastern Marion County in the Ocala National Forest. The sand pine can survive on the deep, infertile sandy soil although it does not reach full size. Because of the large stand of scrub sand pine, the Ocala National Forest is frequently referred to as the Big Scrub.

Florida has a relatively high annual rainfall and during an average year will have precipitation each month. However, there are areas that lack trees because of soil or other conditions rather than the climatic regime. With a knowledge of soil conditions it is possible to make some predictions whether an area is tree-covered or grass-covered in its natural state. The Everglades have a thin veneer of decayed vegetation overlying limestone. Before the building of dikes around Lake Okeechobee and drainage ditches from the lake to the sea, water covered the region for many months of the year—too much in some places to permit trees to grow. As a result, saw grass, a sedge, became the dominant type of vegetation throughout the wetter sections of the Everglades. Such an area is known as **marsh.** In the higher sections pine trees occupied the drier soils, and hardwoods grew in the hammocks.

This pattern holds relatively well throughout the state. Hardwoods dominate the areas where the soil is heavier and moister and pine trees, palmetto, and gallberry dominate the drier, lighter soils.

Thus far, the discussion has differentiated mainly between trees and grasses. This should not be taken to infer that there is a limited number or distribution of species of flora, because Florida hosts a great variety of vegetation species. The cold northwesterly winds of winter have blown seeds from the continent south to the peninsula. In the summer, hurricanes and tropical storms have blown seeds northward from the Caribbean area. Florida has nearly half of all of the species of vegetation native to the United States, and it has many species that do not grow in other states. Most of the unique species are subtropical plants. Some encyclopedias state that the mango is a tropical fruit which is not grown in the United States. People who have lived in South Florida know that many mango trees are found there. Other exotic fruits include guava, papaya, and the avocado. A tropical tree, the Sabal Palm, is the state tree and is on the state seal (Figure 8-2). The Sabal Palm is named for the large leaf bud on the top of the trunk. Many times the tender part of the bulb has been cooked and eaten as a vegetable called swamp cabbage. It is the state tree primarily because of the large number of settlers who were saved from starvation by eating the heart of the palm. Palms are really grasses, not trees. They grow upward and outward from a single terminal bud and do not show annual growth rings as true trees do. However, because of their appearance they are considered trees.

An exception to the tropical species is the Torreya tree (*Torreya taxifola*), an evergreen with shining stiff, narrow needles about an inch long. It belongs to the Yew family and is one of the rarest trees in existence. The Torreya is found only within a 25-mile radius along the banks of the Apalachicola River with its relatives limited to California, China, and Japan. Legend has it that this is the tree, also called gopher, from which Noah built his ark. Some Floridians around Bristol believe that this area was the site of the original Garden of Eden. The Torreya State Park, north of Bristol, is an area set aside to preserve the Torreya tree.

Figure 8-2. The Sabal palm or cabbage palm (Sabal palmetto) is one of more than a dozen species of palm native to Florida. The Sabal palm is the state tree.

IMPORTED SPECIES

Many exotic plants have been brought into Florida by residents and visitors and have adapted well to the state's climate and soil. Some have been beneficial, while others have become such pests in their new environment that it is now a multimillion dollar enterprise to control them.

The chief scourge of Florida's waterways is the water hyacinth (*Eichornia crassipes*) (Figure 8-3). It is native to South America but has spread throughout most of the warm waters of the earth. Seeds were purchased for the New Orleans Cotton Centennial in 1884 from Venezuela by the Japanese government for patrons who attended the Centennial. Two attendees, James and Estelle Chalmers, brought some seeds to Florida. Mr. Chalmers planted the seeds in his lawn fountain. In a short time they had overgrown the fountain and were pulled out and dumped into the St. Johns River. By 1896, the water hyacinth had spread throughout the St. Johns River Valley, and four years later steamboat navigation between Jacksonville and Palatka had virtually ceased because of the spread of the floating tubers. Other gardeners and flower fanciers also aided the spread of the plant.

This beautiful floating plant with its glossy leaves and fragile flowers has cost the taxpayers of Florida millions of dollars annually in attempts to eradicate it. Not only does it clog waterways, making navigation difficult, but it also wastes tremendous quantities of water through evapotranspiration. Approximately 3.5 times more water is lost than that due to surface evaporation for an equivalent area. Eutrophication of lakes is also hastened as the lower leaves die, fall to the bottom and consume oxygen as they decay.

Figure 8-3. Water hyacinths (Eichornia crassipes) are native to South America. Since being introduced into Florida in 1884 by James and Estelle Chalmers, they have become such a nuisance by clogging water bodies that they cost the State of Florida millions of dollars annually to combat them.

Another imported weed that has created problems in Florida's fresh water bodies is the hydrilla. This troublesome submerged weed was introduced as an aquarium plant, but because most aquatic plant dealers grow and harvest the plants in public waters, the hydrilla soon spread to adjacent lakes and streams. Before long, it spread even further to fresh water bodies throughout the region. (In the United States there are no limitations on aquatic plant imports.)

Submerged aquatic plants are much more difficult to eradicate than surface-growing plants. hydrilla is especially difficult to eradicate because, even though the top growth is killed by chemicals injected into the water, the plant produces a large number of hydrosoil turions which sprout as soon as the top growth is killed.

The problems created by the introduction of the water hyacinth and the hydrilla are world-wide in scope and not just limited to Florida. Much more stringent regulations must be enacted and enforced to prevent other troublesome plants from being introduced into the state.

The State of Florida is evaluating a variety of methods to control hydrilla, water hyacinths, alligator weeds, and other troublesome weeds. Alligator weeds have, for many years, been the most prolific native weed to clog Florida waterways. Since 1970 this weed has been successfully controlled biologically through the introduction of foreign flea beetles, stem borer moths, and thrips. Scientists are now working on biologic control of water hyacinths by releasing two imported weevil species and a South American moth—all of which ravenously feed on hyacinths. Manatees also feed on hyacinths, but unfortunately their numbers are declining.

Two biologic approaches are currently under evaluation in hydrilla control. Asian leaf-mining flies and weevils offer promising methods of insect control. White Amur fish have also been imported from Southeast Asia and released in Central Florida lakes. These fish have successfully devoured all hydrilla from the lakes in which they have been introduced. However, these fish are unable to reproduce in our lakes due to improper breeding conditions. Hybrids of amur-bass are being developed to solve this problem.

VEGETATION TYPES

The native vegetation of Florida is so diverse that it is impractical in a geography text to describe or even mention all of the species that grow within the boundaries of the state. More than 300 species of trees grow in Florida, and as many as 850 species of plants have been identified in south Florida alone.

Since geographers are primarily interested in the distribution of vegetation and its relation to the physical conditions and cultural influences of the region, the authors have used Davis' Vegetation Types of Florida as a basis for describing the native vegetation of the state (Davis, 1967) See Figure 8-4.

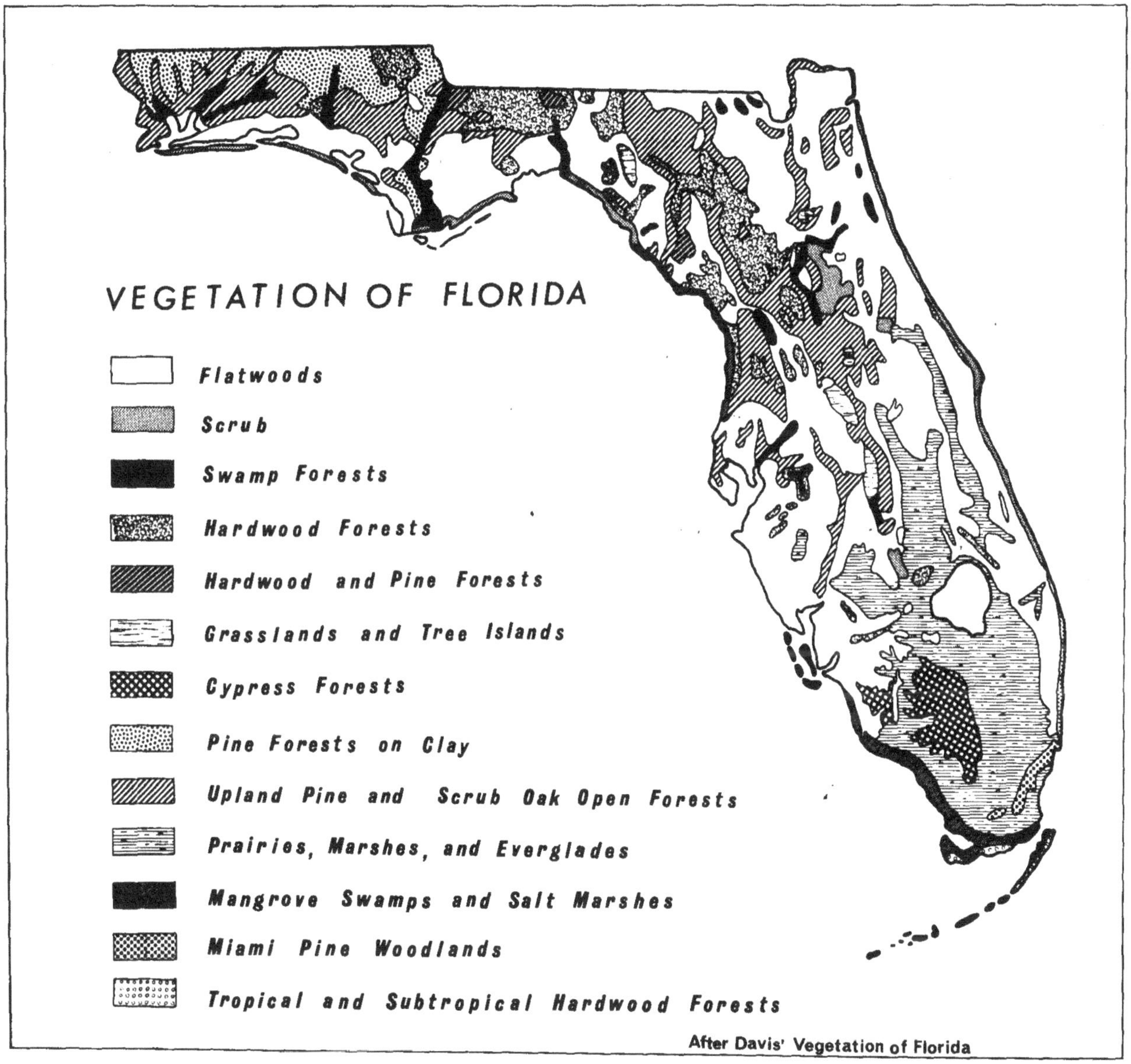

NATIVE VEGETATION TYPES OF FLORIDA

I. The Flatwoods
II. The Scrub
III. The Swamp Forests
IV. The Hardwood Forests
V. The Hardwoods and Pine Forests Regions
VI. The Grasslands and Tree Islands Region
VII. The Cypress Forests
VIII. The Pine Forests on Clay
IX. The Upland Pine and Scrub Oak Open Forests
X. The Prairies, Marshes and Everglades Region
XI. The Mangrove Swamps and Salt Marshes
XII. The Miami Pine Woodland Region
XIII. The Tropical and Subtropical Hardwood Forests

The Flatwoods

Slightly less than 40 percent of the area of the state is covered with pine flatwoods. The area in which the flatwoods predominate roughly coincide with the coastal plains on either coast, but also includes sections of the central highlands and the flood plain of the Kissimmee River before it was straightened. Beginning just south of the St. Marys River in northeast Florida, the flatwoods extend south along the coast to the approximate latitude of Miami. On the Gulf Coast this type of vegetation is characteristic of the areas reaching from Escambia County east and south to northern Collier County. The most extensive areas are found in northeast Florida around Baker, Bradford, and Nassau Counties; in Wakulla and Taylor Counties along the northern Gulf Coast; and in central Florida in the vicinity of Manatee, Sarasota, Hardee, and DeSoto Counties. All of the counties listed, except for the last four, have over 75 percent of the area of the respective counties in open woodland type of forest.

Throughout the entire region, the topography is predominantly flat with elevations of usually less than 100 feet. The drainage varies, but usually the soil is poorly drained with a high water table. In most sections of the region, the soil is classified as Spodosol overlying non-calcareous materials. In lower areas and depressions, peats and mucks are present. The latter areas frequently contain ponds and swamps and support bald cypress, water tupelo, red maple, water hickory, and cabbage palm. Some hammock areas are found scattered throughout the flatwoods. These are especially numerous in Hillsborough and Polk Counties.

Longleaf, slash, and pond pines are the dominant species of trees in the flatwoods. Longleaf pine is usually found on the drier soil, while the more poorly-drained areas are covered with slash pine, and in some areas, pond pine. The density of forest cover varies throughout the region depending upon the frequency of forest fires and the type of logging activities conducted. Pine trees need abundant sunlight and will not grow well under a canopy of taller trees. Likewise, the thickness of the undergrowth varies with the density of the forest cover.

The marshes along the St. Johns River are practically devoid of trees, but farther from the river, the pine trees are more abundant. The thickest stands of pine trees are located in the northern sections of the flatwoods, with the trees being more widely dispersed the farther south they grow. Saw palmetto and wiregrass are the two most widely distributed undergrowth plants, but gallberry, myrtle, and rosemary are also characteristic of the region.

The Scrub

Florida's scrub occurs in several separate areas, the two largest located within or near the central highlands. The largest area, occupying over a quarter of a million acres, is part of the Ocala National Forest. Most of the forest lies in Marion County except for the eastern edge which extends into Putnam and Lake Counties. In 1908, President Theodore Roosevelt set aside the land for the Ocala National Forest principally because the soil was so poor that the land at the time seemed good for little else. Within the boundaries of the forest is the largest area of scrub pine (*Pinus clausa*) or sand pine in the world. The sand pine will grow on the deep sands of the Citronelle formation which covers a large portion of the region.

Scattered throughout the Big Scrub are slight depressions covered with herbaceous plants and rimmed by slash pine. On the better soils are found longleaf, slash, loblolly, and scrub pine; in the wetter areas cypress and hardwoods frequently mix with cabbage palms. The United States Forest Service selectively cuts the trees of commercial value and replants with the most valuable trees that the particular soil of the area will support.

A second large area of scrub forest is located at the southern tip of the highlands ridge in the vicinity of Lake Stearns. The outstanding characteristic of this section is the great variety of low trees and shrubs with some dense stands of sand pine (Figure 8-5), varying in size from small to medium. Some areas are devoid of sand pine but contain slash or longleaf pine among the woody shrubs. The scrub oaks in this region consist of myrtle oak, chapman oak and scrub live oak. Florida hickory is characteristic of the highlands ridge but may be non-existent in other scrub areas. Undergrowth usually contains rosemary, and on the surface of the dry sands, lichens and prickly cactus fight for survival.

Figure 8-5. A hardwood swamp forest in northeast Florida. Streams overflow their banks during the wet season and deposit clay along the banks. The clay will support hardwood trees with some saw palmetto undergrowth.

Other smaller areas with scrub trees include the dunes paralleling the Atlantic coast between Fort Lauderdale and Fernandina Beach and along the southwest coast from Fort Meyers to Cape Romano, including Marco Island, south of Naples. Some dune scrub occurs from Panama City to Pensacola.

The Swamp Forests

Most of the swamp forests of Florida are located on the bottomlands adjacent to streams and are covered with water during a large part of the year. The streams of northwest Florida are larger and muddier than the streams of south Florida; thus, the most extensive area of swamp forest are found in northwest and northeast Florida. The largest section covered with swamp forest is along the Apalachicola River and extends from the mouth of the river almost to the Georgia boundary. Smaller areas are found along the Aucilla and Santa Fe Rivers.

In southern Florida the larger swamps are located along the Peace, Taylor, and Fisheating Creeks. Smaller ones occur along sections of the Caloosahatchee, Turner, New, and St. Lucie Rivers. The characteristic swamp vegetation along the creeks, rivers, or sloughs are red maple, water oak, and laurel oak. Florida ash, in addition to the ever-present bald cypress. Wax myrtle, sweet bay, red bay, cabbage palm, and strangler fig are among the plants found in the southern Florida Swamp Forest.

The swamp forests along the Apalachicola River and other rivers of northwest and northeast Florida usually contain red maple, black gum, tupelo gum, ash, water oak, and elm. Cypress is almost always found among the hardwoods.

The Hardwood Forests

Two areas in northeast and northwest Florida constitute the bulk of the hardwood forests, with smaller islands of land supporting this type of vegetation which extends from eastern Walton County along the Choctawhatchee River to northeastern Glades County near Lake Okeechobee. The largest area is located entirely within Alachua and Marion Counties, on the phosphatic limestone soils, with smaller outliers extending north across the Santa Fe River into Columbia, Union and Bradford Counties and paralleling the West side of the Suwannee River from Branford almost to the Gulf of Mexico. The second largest areas lies along the Florida-Georgia boundary between the Apalachicola and the Aucilla Rivers. A third area of considerable size is in Jackson County in the Marianna Lowlands. The soils throughout the northern region are Ultisols, consisting of red clays and loams in the Marianna Lowlands and the northern highlands and poorly-drained sands and loamy sands (Entisols) north of Lake Okeechobee. However, most of the soils throughout the region are underlain by some type of calcareous deposits with the surface materials containing clay and sandy clay.

On the calcareous and phosphatic hammock land of Alachua and Marion Counties are found red oak, laurel oak, swamp chestnut, hickory, dogwood, and sweet gum, with scattered cabbage palms. Spanish moss is abundant on the trees in this area.

Along the Georgia border in the northern highlands and the Marianna Lowlands, some short leaf, loblolly, and longleaf pines grow on the drier uplands while the hardwoods are found on the slopes and on the heavier soils. These include red oak, white oak, dogwood, redbud, sweet gum, bay, and black gum. The scarcity of new palmetto and wiregrass undergrowth is noticeable in this area.

Hardwood and Pine Forests

There are no extensive areas of hardwood and pine forests; instead, several small islands of hardwood and pine forests lie adjacent to the land covered by hardwood forests or by upland pine and scrub oaks.

On the eastern side of the Apalachicola River in northwestern Gadsden County, the land slopes steeply toward the river and in several places forms precipitous slopes. Everywhere the topography is hilly with deep rich alluvial soils in the valley. On the slopes short leaf pines are mixed with magnolia and other hardwoods. Beech, bay, poplar, and water oak tend to be limited to the ravines, while magnolia, holly, willow, maple, and white oak grow both on the bluffs and in the ravines.

In northern Jefferson County there is another oval-shaped area containing pine and hardwood trees. The terrain is hilly with low, gentle slopes characteristic of the Tallahassee Hills section of the northern highlands. On the drier uplands the original forest cover was short leaf pine, red oak, hickory, dogwood, redbud, and white oak. Much of the pine today is second growth. Longleaf pine grows on the sandier soils but differs from the flatwoods in the scarcity of wiregrass and saw palmetto undergrowth.

The southernmost of the three areas included in the hardwood and pine forests is located on the eastern side of the Suwannee River, just south of Manatee Springs State Park in Levy County. The dominant hardwood is the red oak with longleaf and loblolly pines.

Grasslands and Tree Islands Region

The areas of Florida that are classified as the grasslands and tree islands region are covered with surface water during a part of the year and are dominated by grasses and aquatic plants in the pools and deeper bodies of water. These areas differ from marshes in that the depth of the water over the ground surfaces is usually shallower and does not stand for as long a period of time during the year, and the dominant vegetation is grass with some swamp areas containing trees, usually as islands of forest vegetation (Figure 8-6).

Figure 8-6. *Tree Island located within the Lake Istokpoga-Indian Prairie Basin north of Lake Okeechobee. The clusters of palms break the monotony of the vast area of flat grasslands.*

One of the most extensive areas of the grasslands is located within the Lake Istokpoga-Indian Prairie Basin, which lies along the eastern side of the highland ridge south of Sebring. Lake Istokpoga, at the head of the basin, overflows across nearly flat and poorly drained peats and mucks to the south. The swamp and marsh vegetation growing on the peats and mucks, south of Lake Istokpoga, separate the lake from Indian Prairie, a vast level prairie dotted with cabbage palm islands. The predominant vegetation of the prairie is switch grass. Canals have been dug to drain the excess water from Indian Prairie.

The Cypress Forests

Many of the wet soils of Florida contain cypress mixed with hardwoods and other water-loving trees. In southern Florida there are many areas around sloughs, fresh water lagoons, and shallow ponds that are almost completely dominated by cypress. These areas are known as Telegraph Cypress in Charlotte County and Six-Mile Cypress in Lee County. There are several cypress swamps in Hendry and Collier Counties and along the Loxahatchee Slough in Palm Beach County.

The density of the cypress cover throughout the region varies from dense swamps along the streams and other bodies of water to open cypress forests. in the Big Cypress region. Within Big Cypress there are many cypress **domes** which consist of strands of tall cypress trees in the center of the area and shorter cypress trees growing out from the center. Frequently an open pond or body of water occupies the center of the dome.

Along the sloughs and other linear bodies of water, cypress trees grow in elongated patterns called **strands** (Figure 8-7). Frequently, the strands extend out from the cypress domes. Both the domes and the strands support bay trees, wax myrtle, cocoaplum, and other plants as well as a preponderance of cypress trees.

Figure 8-7. Cypress dome located between Otter Creek and Bronson. The road divides the cypress dome in half and the water which once formed the center of the dome was filled in to build the road. In a typical cypress dome, the trees nearest the water are tallest and the height of the trees decreases as the distance from the water increases.

In Big Cypress Swamp and other parts of southern Florida, there are areas covered with scrub cypress which resemble the open pine sections of the flatwoods. Usually such areas consist of prairies and marshland areas or wet grassland. The largest areas of scrub cypress are located in eastern Collier County and northern Monroe County, adjacent to the Everglades.

Pine Forests on Clay

The pine forests on clay lie west of the Apalachicola River and almost entirely within the Marianna Lowlands and the Northern Highlands. One smaller area lies adjacent to the swamp forest along the western side of the Apalachicola River near its mouth.

Originally, open forests of longleaf pine covered a large part of the region. Slash pine occupied the wetter soils and the areas close to shallow ponds and were frequently mixed with cypress and black gum trees along the edge of the ponds. In the northern part of the Marianna Lowlands, the wiregrass, saw palmetto, and gallberry undergrowth was much less dense than in the coastal regions. This was mainly due to repeated burning and frequent forest fires.

Much of the timber shipped from the Port of Pensacola during the period 1876 to 1907 was logged in this region. Slash pine has replaced the longleaf pine in the stands planted for pulpwood because it grows faster than longleaf when protected from fire.

The vegetation of the pine forests is similar to that of the flatwoods. The soil, however, is quite different, being Ultisols instead of the Spodosols of the flatwoods. Throughout the region the underlying profile of the Ultisols consists of clay or clay loam with a thin veneer of sand on the surface. The clays and sands of the Marianna Lowlands are underlain by limestone which is exposed along some of the streams.

Upland Pine and Scrub Oak Open Forests

The upland pine and scrub oak open forests extend from Escambia County in northwest Florida east to the central highlands and follow this physiographic region south to its southern limit near Sebring. Except for small areas in Hernando and Pasco Counties, in Pinellas and Hillsborough Counties, and along the Peace River in DeSoto and Hardee Counties, most of this type of vegetation growing on the peninsula is confined to the central highlands.

The Entisols of this region consist of excessively to moderately well-drained sands which are typical of the sandhill sections of the central highland, especially in the lake region. Originally, the soils of the area supported open forests of longleaf pine with the driest sections of the upland containing black jack and turkey oaks. After the longleaf pines were logged, turkey oaks and black jack oaks increased in number and size until numerous areas today look like scrub oak forests. In many sections of the region trees have been completely removed and the land planted in citrus, as the soil is well-drained and the rolling topography provides citrus trees with the necessary air drainage during cold spells.

Scattered throughout the entire region are shallow ponds and sinks. In the northern part of the state cypress, slash pine, or black gum trees grow near the shallow bodies of water. In most of the limestone section of the central highlands the sinks are treeless. There are trees, however, along the swamps bordering the Suwannee River.

The turkey and black jack oak trees are of little commercial value. Even in areas where they form the majority of trees, the density is not great enough to profitably remove them for pulpwood. Some have been cut to make charcoal for outdoor grills, but production has not been extensive. The undergrowth found in the longleaf pine and turkey oak forests is not as dense as in the flatwoods. Saw palmetto is less common than it is among the pines of the flatwoods. Wiregrass grows well, as does myrtle and rosemary, the latter growing on the poorest soils.

The Prairies, Marshes, and Everglades

Marshes are defined as areas covered with water at least part of the year which support a dominant vegetation of grasses, sedges or reeds. Marshes differ from swamps in that trees are the dominant type of vegetation in swamps; grasses predominate in marshes (Figure 8-8).

Figure 8-8. Closeup of sawgrass (Mariscus jamaicensis). Although termed a grass, the plant is actually a sedge. The serrated edges of the sawgrass plants are sharp and can inflict painful cuts on a person who attempts to walk through the tall grass.

Payne's Prairie, south of Gainesville in Alachua County, is an excellent example of a prairie that occupies the bottom of an intermittent lake draining into a sink. Before the construction of drainage ditches across the prairie, water would often accumulate during periods of heavy precipitation or when the Alachua Sink became clogged. During the 1880s, the area was covered with water, and boats hauled citrus from one side of the lake to the other. The flat terrain supports a dense cover of grass and, in the wetter portions, aquatic plants including the water hyacinth. The Indians used the area to graze cattle before it became privately owned. It is now a state-owned wildlife refuge.

Most of the areas of Florida containing marshes are located south of the 28th parallel (line across the state from Tampa to just south of Melbourne) with the largest continuous areas being in the Everglades, covering some 2.5 million acres. The vast, flat, former sea bottom of the Pamlico Sea consists of marshes and prairies interrupted by a scattering of islands of bay trees and hammocks. Also included with the Everglades are ponds and sloughs containing a variety of aquatic life.

Saw grass (*Mariscus jamaicensis*) dominates the area of the Everglades underlain by peat and muck soils. In parts of the Everglades National Park, the "River of Grass," (Pa-hay-okee) stretches across the horizon as far as the eye can see. Actually, the saw grass is a sedge and not a grass. Estimates place the approximate size of the Everglades at one million acres with varying densities of saw grass cover.

The custard apple and the willow-elderberry swamps formerly made up a wide band of vegetation around the southern edge of Lake Okeechobee from Canal Point to Moore Haven. Most of the area has been cleared, and the peats and mucks are used for agricultural purposes (Figure 8-9).

Figure 8-9. The native vegetation of the Everglades, along the southeastern shore of Lake Okeechobee (Pahokee to Belle Glade), was originally custard apple and elderberry as shown in this picture taken near Belle Glade before the area was cleared.

Throughout the Everglades are sloughs and other bodies of water containing aquatic vegetation, and on mounds there are islands of trees. The aquatic vegetation consists of floating and submerged plants, such as water lilies, pond weeds, bladderworts, and numerous grasses. Red bay, sweet bay, wax myrtle, and daboon holly usually constitute the bay-head type of tree island in the region with several species of ferns forming the undergrowth. Bordering the Everglades area are several types of marshes which frequently form the transition zones. These include flag marshes dominated by blue flags, arrowhead and non-grass herbs; cattail marshes often occupying brackish waters; and spike rush marshes consisting predominantly of spike rush with lesser numbers of bulrushes, fringe rushes, maiden cane, and panic grasses.

North of Lake Okeechobee two areas of marsh occur. The larger one contains the flood plain of the Kissimmee River with its alluvial soil formed (before becoming a channel) from repeated inundations during high water. Grasses, sedges, aquatic plants, vines, and shrubs with a few willows make up the vegetation pattern of this area.

Around the lakes at the headwaters of the St. Johns River in southern Brevard County is another area of marshes. The peat and muck soils are covered with varying depths of water during the year and support reeds, floating leaved aquatic plants, and, in some sections, stands of saw grass.

Mangrove Swamps and Salt Marshes

The mangrove trees are unique among the trees of the world in that they can and do grow in salt water. Actually there are three species of mangrove trees in Florida: red, black and white. All three can live in salt water. Along the southwestern coast the mangrove forests have developed into zones. The outer zone along the shore is occupied by the red mangrove with its characteristic prop roots giving it the appearance of a tree walking. The black mangrove lives a little further inland from the shore on the mud flats and produces many erect roots (pneumatophores) which stick up through the mud. The third zone, which is most inland, is occupied by the less common white mangrove with its oval-shaped leaves which are alike on both sides. The inner zone usually contains buttonwood trees among the white mangroves.

Mangrove forests extend from the lagoons south of St. Augustine on the coast of northeast Florida south around the southern tip of the peninsula and as far north as Cedar Key on the Gulf coast. Killing frosts limit the northern extension of mangrove trees. The main section of this region in Florida is along the southwestern coast from dense forests of tall trees decreasing in height further inland. The belt narrows in the vicinity of Chokoloskee Bay and comes to a point around Naples.

In 1960, Hurricane Donna killed many of the mangrove forests around Flamingo and along the Gulf Coast (Figure 8-10). A sheet of ocean water several miles wide and approximately 20 feet high was blown out of Florida Bay by winds in excess of 130 knots (150 mph). With this water came a fine, impervious marl mud that was deposited on the forest floor and probably caused an oxygen deficiency in the root hairs of the completely defoliated trees (Craighead, 1969). It was only in the low areas, less than a foot above sea level, that the majority of the trees were killed.

Figure 8-10. When Hurricane Donna came through Florida Bay and the Everglades in 1960, it killed approximately 25 percent of the mangrove trees. In 1960 the trees had just about recovered from the Labor Day hurricane of 1935. This picture shows the dead mangroves near Flamingo two years after the hurricane had passed through.

Salt marshes may occupy the transition zone between the mangrove trees of the salt and brackish waters and the saw grass which lives in fresh water. Because the water in the transition zone is often alternately fresh and saline, plants growing in the zone may be mixed. Among the plants that are saltwater tolerant are salt, cord, and bunch switch grasses. Cape Sable has an area of salt flats which is not usually wet. The plants here are similar to other salt marshes but contain open bare areas of high salt concentration.

Miami Pine Woodlands Region

The Miami pine woodlands region occupies roughly 500 square miles of the upland sections of the Miami oolitic limestone ridge, separating the Everglades from the coastal strip in Dade County. Throughout the drier parts of the region, the slash pine (*Pinus efliottil densa*) was originally the most widely distributed tree and resembled the slash pine of other areas of Florida, except for being smaller in size. This resulted mainly from the lack of soil found on the Miami Rock Ridge. Much of the land formerly covered with slash pine has been cleared for agricultural or urban use.

The Tropical and Subtropical Hardwood Forests

The tropical and subtropical hardwood forests are along the southern coast of the peninsula and on the Florida Keys. In the southern hardwood forests the trees and shrubs are subtropical with a gradual transition to tropical species southward, the only completely tropical forests being in the Keys.

One of the outstanding characteristics of the hardwood forests of southern Florida is the variety of plants found in the hammocks. As many as 40 to 50 species of large trees, small trees, and shrubs combine to form dense undergrowth in the subtropical hammocks. Among the typical tropical and subtropical trees are the strangler fig, wild tamarind, gumbo limbo, mastic, bustic, several kinds of stoppers, and the royal palm. Other more northern trees usually found growing in the subtropical hammocks include hackberry, mudberry, and red maple. Some of the common plants are lancewood, ironwood, wild coffee, crabwood, potato tree, myrsine, poisonwood, satinleaf, papaya, pigeon plum, and the Jamaica dogwood.

The abundance of vines, ferns, and epiphytes (plants growing on other plants) is another characteristic of subtropical hammocks. Such ephiphytes as airplants, rare non-terrestrial orchids, and ferns are among those growing in the hammocks. Included with the common vines are rattan vine, bullace grape, possum grape, poison grape, poison ivy, pepper vine, and rubber vine.

Along the southern coast and on the Keys some of the rarest tropical plants in the continental United States are found. These include the West Indian mahogany, the lignum vitae, the thatch palm, and the manchineel. Because most of the Keys have experienced frost, some of the plants are occasionally killed during the severe cold spells.

REFERENCES

Alexander, T.R. (1955), "Observations on the Ecology of the Low hammocks of Southern Florida. *Florida Academy of Science Quarterly Journal.* Vol. 18, pp. 21-27.

———. (1967), "Effects of Hurricane Betsy on the Southeastern Everglades." *Florida Academy of Science Quarterly Journal,* Vol. 30.

Brown, T.W. (1963), *The Ecology of Cypress Heads in Northcentral Florida,* Unpublished Master's Thesis, University of Florida, Gainesville, Florida, 90 pp.

Craighead, F.C., Sr. (1969), "Vegetation and Recent Sedimentation in the Everglades National Park," *Florida Naturalist,* Vol. 42, No. 4.

Daubenmire, R. (1978), *Plant Geography,* Academic Press, New York, 333 pp.

Davis, J.H., Jr. (1943), *The Natural Features of Southern Florida,* Florida Geological Survey Bulletin No. 25, Florida Geological Survey, Tallahassee, Florida.

———.(1940), "The Ecology and Geological Role of Mangroves in Florida," *Carnegie Institute Publication* 517, Washington, D.C., pp. 303-412.

———. (1967), *General Map of Natural Vegetation of Florida* Agricultural Experiment Stations Institute of Food and Agricultural Sciences, Gainesville, Florida.

———. (1971), *The Trees of South Florida,* Vol. 1, University of Miami Press, Coral Gables, Florida, 212 pp.

Engler, F.E. (1950). "Southeast Saline Everglades Vegetation, Florida and Its Management," *Vegetation,* Vol. 3, pp. 213-265.

Harrison, D.S., R.D. Blackburn, L.W. Weldon, J.R. Orsenigo, and G.F. Ryan (1965), *Florida Agricultural Extension Bulletin No. 219B,* University of Florida Agricultural Extension Service, Gainesville, Florida.

Johnson, L. (1974), *Beyond the Fourth Generation,* University of Florida Press, Gainesville, Florida, 230 pp.

Kurz, H. and R.K. Godfrey, with collaboration of K.E. Wagner (1962), *Trees of Northern Florida,* University of Florida Press, Gainesville, Florida.

Marcus, R.B. (1964), *A Geography of Florida* William C. Brown Book Company, Dubuque, Iowa, 162 pp.

———. and E.A. Fernald (1975), *Florida: A Geographical Approach,* Kendall/Hunt Publishing Company, Dubuque, Iowa, 302 pp.

Morton, J.F. (1977), *Wild Plants for Survival in South Florida,* Fairchild Tropical Gárden, Miami, Florida, 80 PP.

Murrill, WA. (1945), *A Guide to Florida Plants,* Rose Printing Company, Tallahassee, Florida, 90 pp.

Vernon, R.O. (1974), "Cypress Domes," *Science,* 105 (2717).

SUMMARY OF CHAPTER 8

Florida, with its temperate to subtropical climate, its seasonal rainfall, and its many different types of soil, supports one of the largest varieties of native vegetation of any state in the union. Forests cover slightly less than one-half of the total land area of the state. The longleaf pine was at one time the dominant needleleaf tree on the drier, sandy soils, but it has yielded to the slash pine which grows faster when protected from fire. Weed trees, such as the black jack and turkey oak, have rapidly increased in number with the advent of forest fire protection. Hardwood trees grow throughout the state, usually on islands of heavier clay soils known as hammocks.

In the Everglades, trees give way to grasses because water covers the ground most of the year. The dominant type of vegetation in the northern and central Everglades is the saw grass which thrives in the water-covered peats and mucks. To the west of the Everglades is the Big Cypress Swamp which is protected from the open salt and brackish waters by mangrove swamps along the southwestern coast.

In addition to the native species of vegetation there are many species that have been imported into the state. Some have been beneficial, but two—the water hyacinth and the hydrilla—have become pests that are extremely difficult to control; clog the streams and waterways; increase the rate of evapotranspiration; and hasten the eutrophication of lakes and ponds.

REVIEW OF CHAPTER 8

Part A—Map Exercise—Vegetation of Florida

In the Appendix there is a blank vegetation map of Florida. Use the following legend to complete the map.

Vegetation Type	Color	Symbol
Flatwoods,	White	Solid
Scrub	Black	Solid
Swamp Forests	Blue	Solid
Hardwood Forests	Brown	Solid
Hardwood and Pine Forests	Brown	/////
Grasslands and Tree Islands	Orange	Solid
Cypress Forests	Blue	/////
Pine Forest of Clay	Green	Solid
Upland Pine and Scrub Oak Open Forests	Red	Solid
Prairies, Marshes and Everglades	Yellow	Solid
Mangrove Swamps and Salt Marshes	Purple	Solid
Miami Pine Woodlands	Green	/////
Tropical and Subtropical Hardwood Forests	Orange	/////

Part B

1. In general, ________________ must be available continuously for trees to grow, while a winter dry season of 2-3 months will result in ________________ as the primary vegetation.
2. The vegetation type that represents a climax for much of Florida is ________________.
3. Small areas of clay and organic soils scattered through sandy soils, and which support hardwood islands are called ________________.
4. ________________ is recognized as the state tree of Florida, although it is really a grass.
5. The imported exotic water plant which is most difficult to eradicate is ____________.
   a) Torreya tree
   b) water hyacinth
   c) hydrilla
   d) magnolia

Appendix

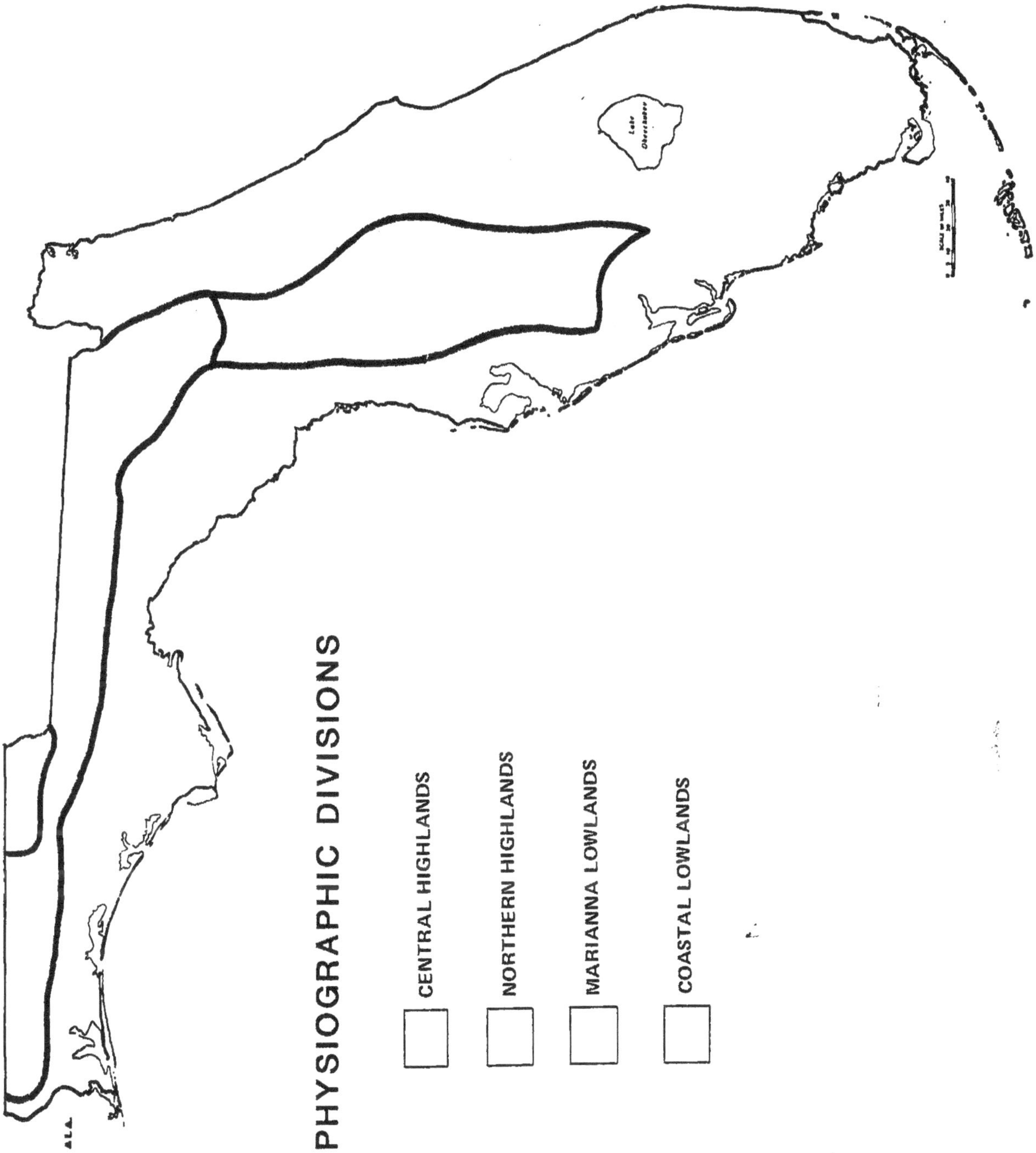
PHYSIOGRAPHIC DIVISIONS
CENTRAL HIGHLANDS
NORTHERN HIGHLANDS
MARIANNA LOWLANDS
COASTAL LOWLANDS

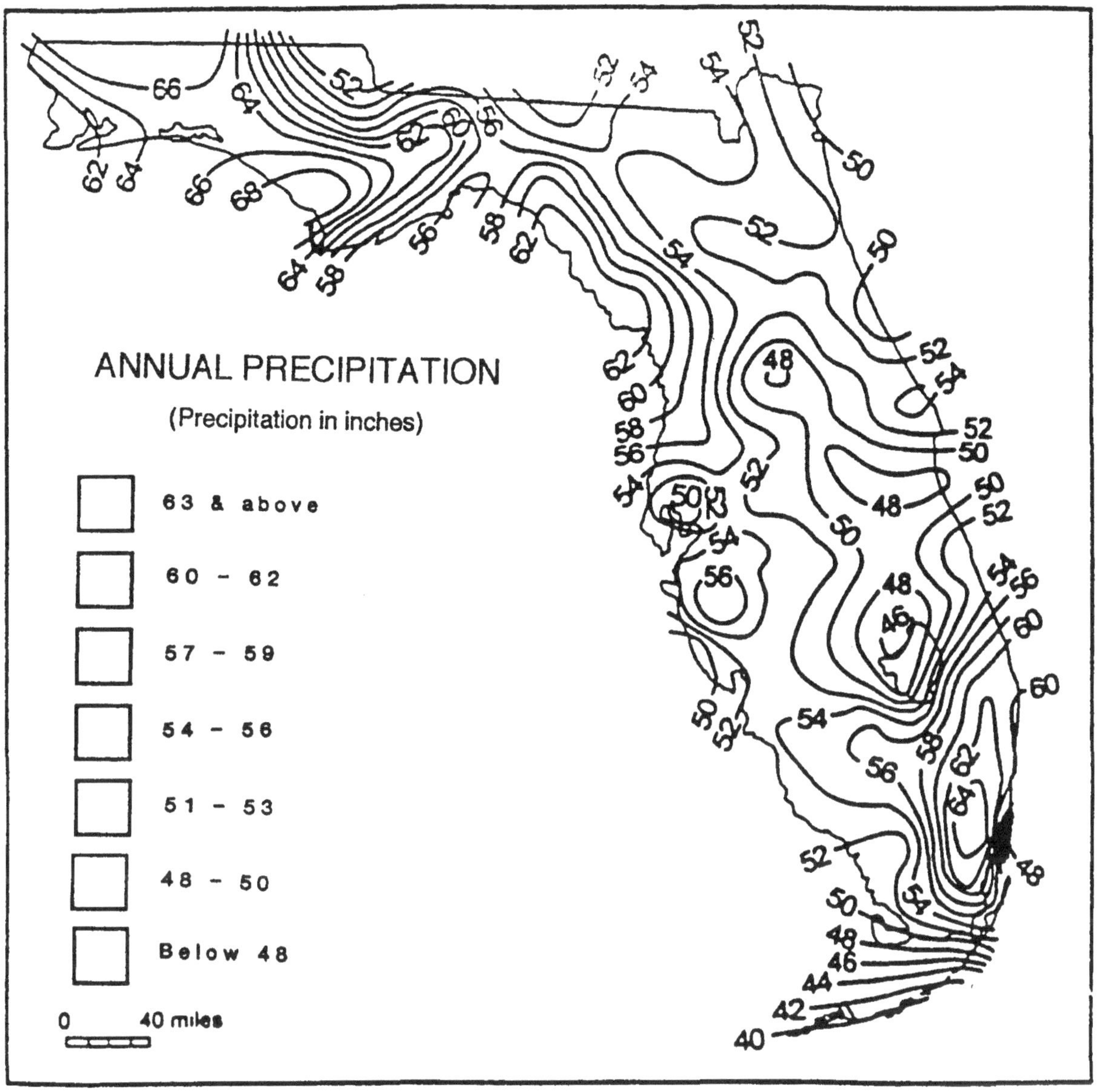
ANNUAL PRECIPITATION
(Precipitation in inches)
63 & above
60 - 62
57 - 59
54 - 56
51 - 53
48 - 50
Below 48
0 40 miles

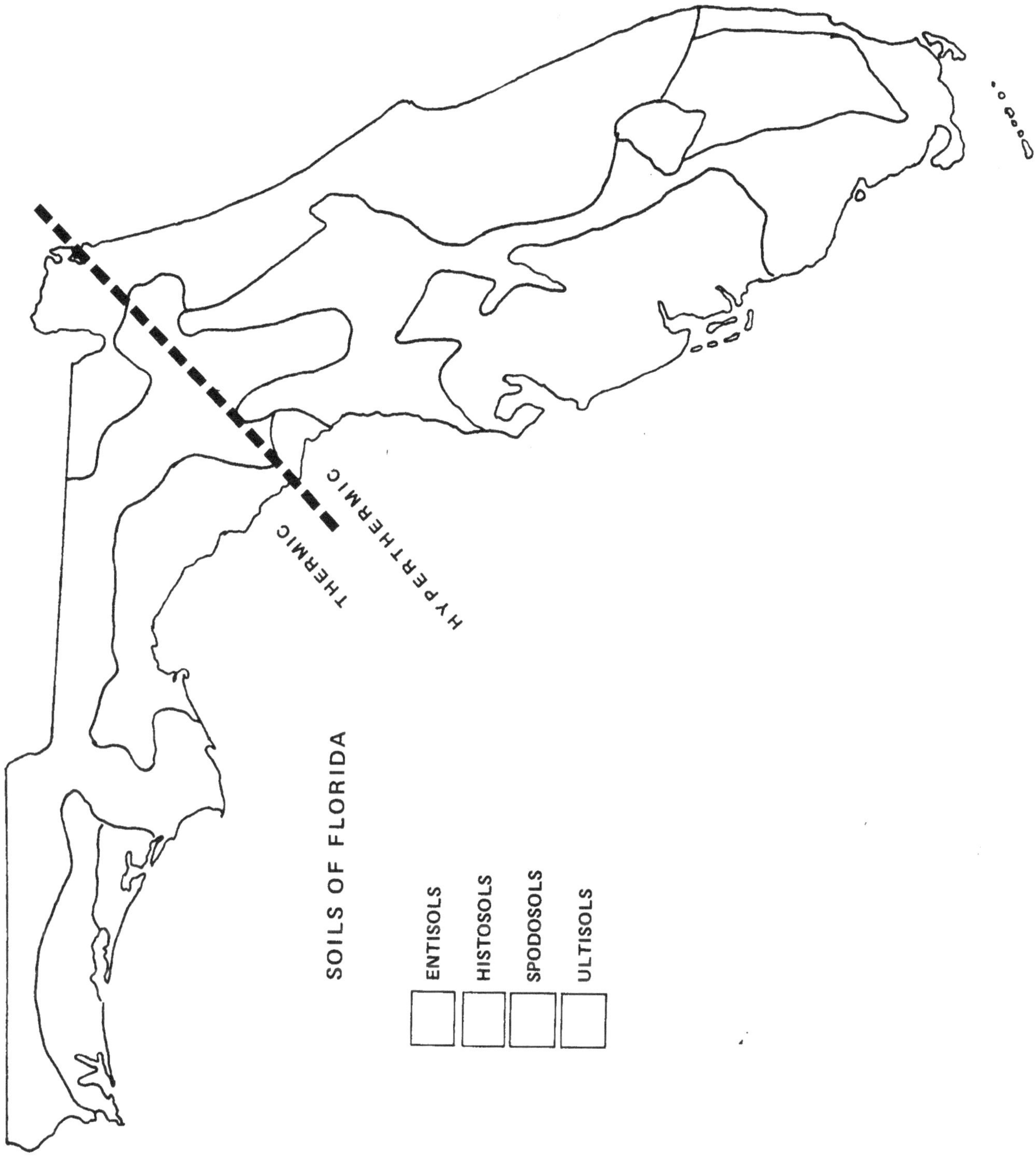
THERMIC
HYPERTHERMIC
SOILS OF FLORIDA
ENTISOLS
HISTOSOLS
SPODOSOLS
ULTISOLS

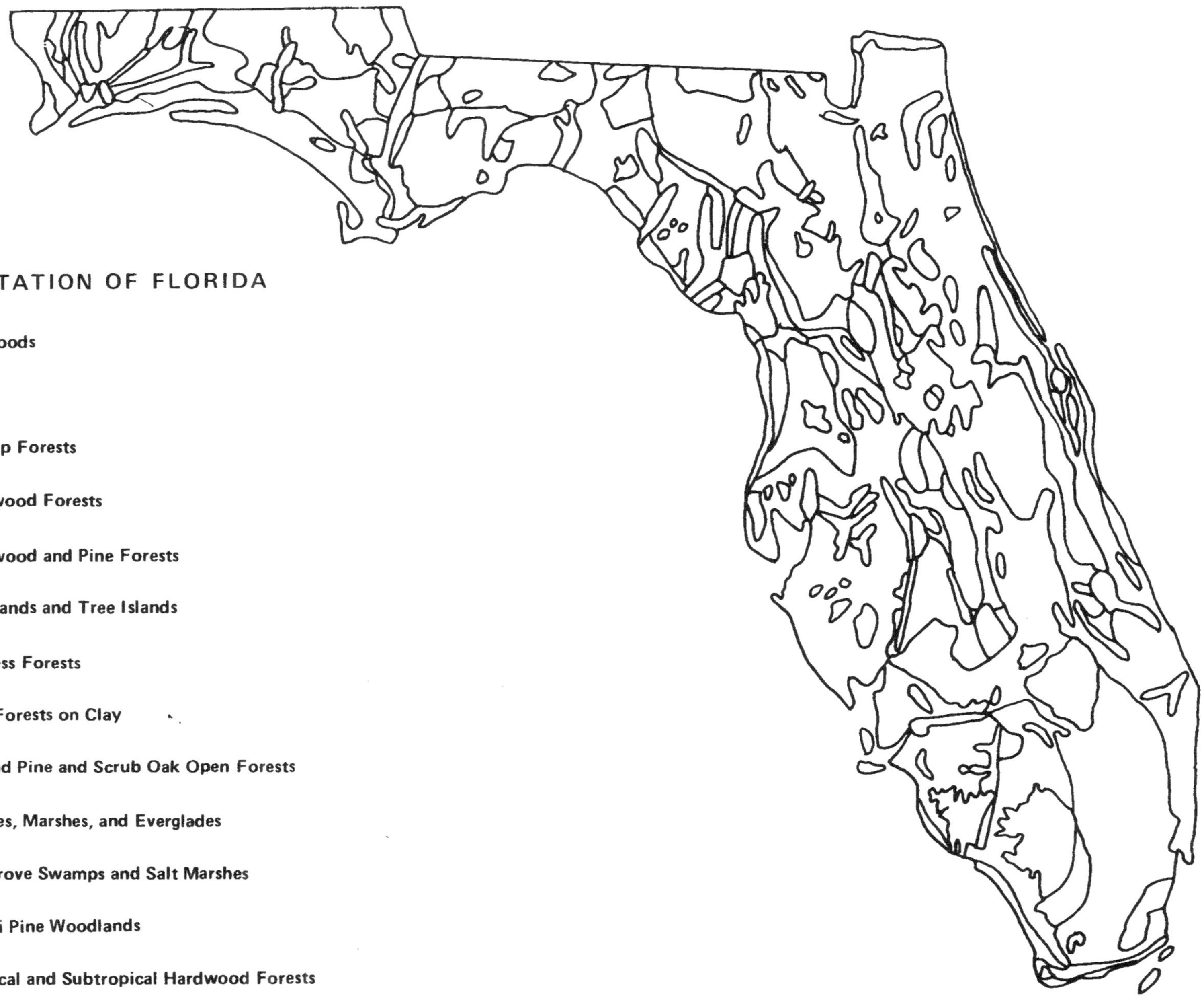

VEGETATION OF FLORIDA

- Flatwoods
- Scrub
- Swamp Forests
- Hardwood Forests
- Hardwood and Pine Forests
- Grasslands and Tree Islands
- Cypress Forests
- Pine Forests on Clay
- Upland Pine and Scrub Oak Open Forests
- Prairies, Marshes, and Everglades
- Mangrove Swamps and Salt Marshes
- Miami Pine Woodlands
- Tropical and Subtropical Hardwood Forests

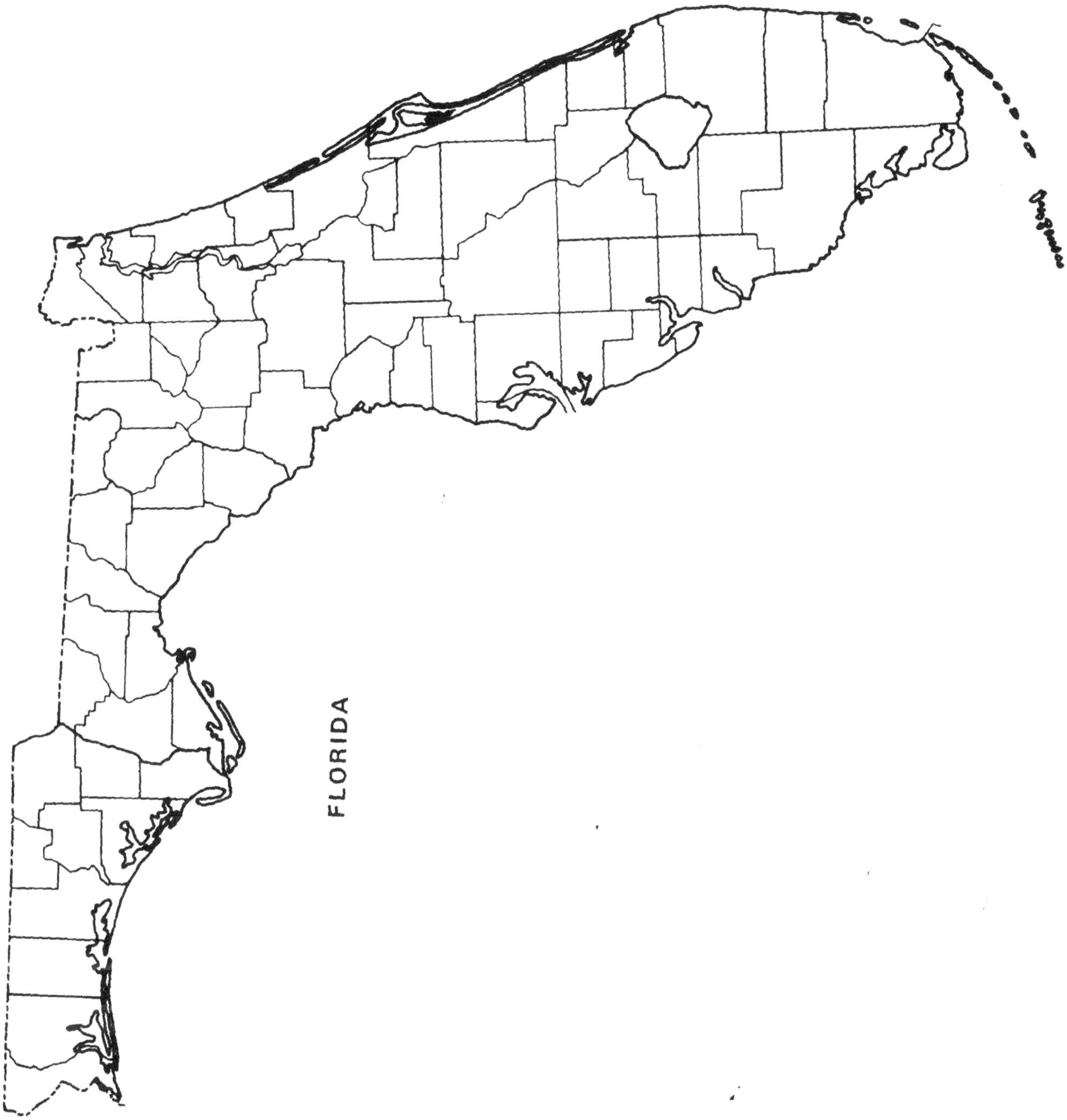
FLORIDA

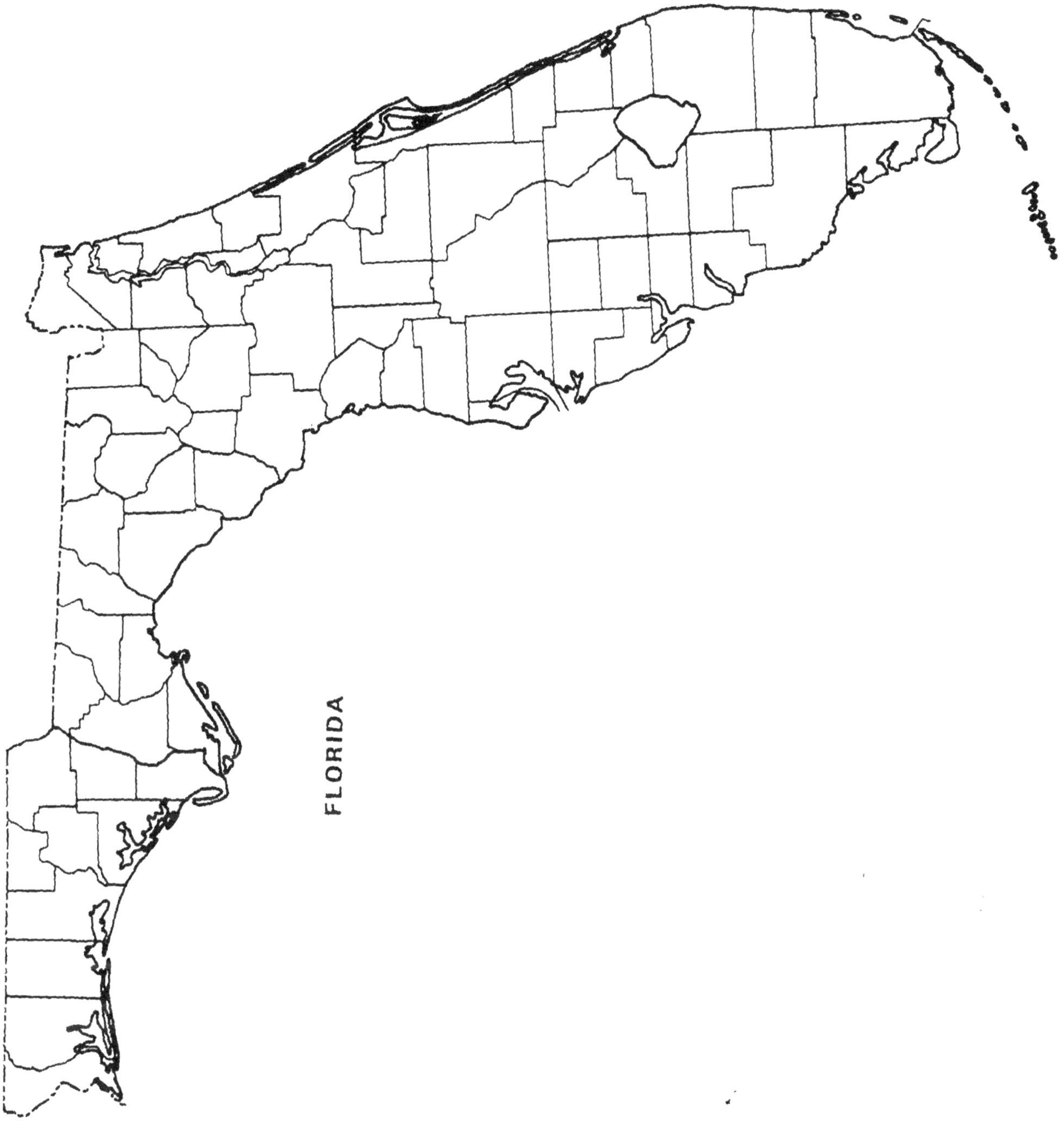
FLORIDA

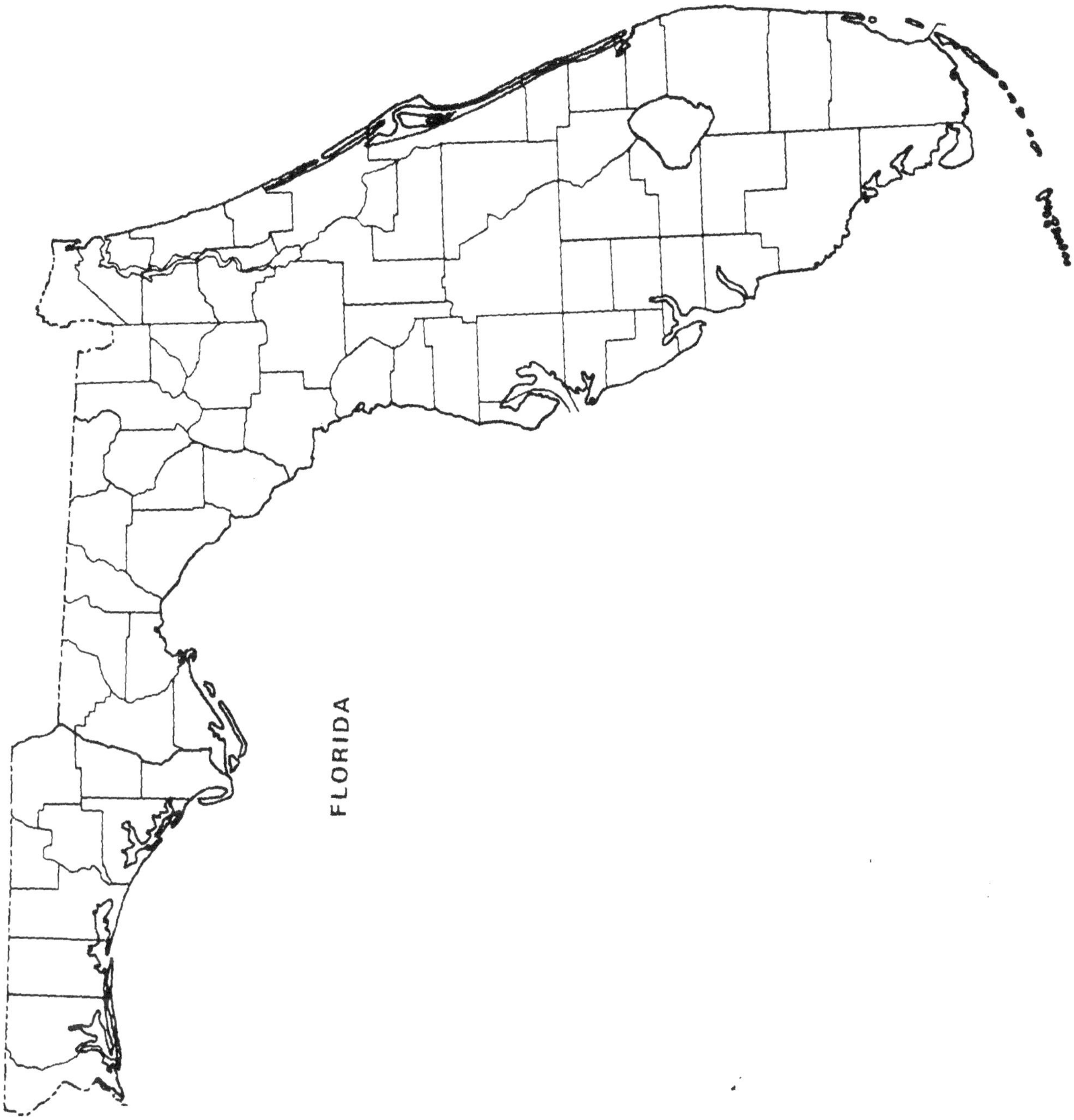
FLORIDA

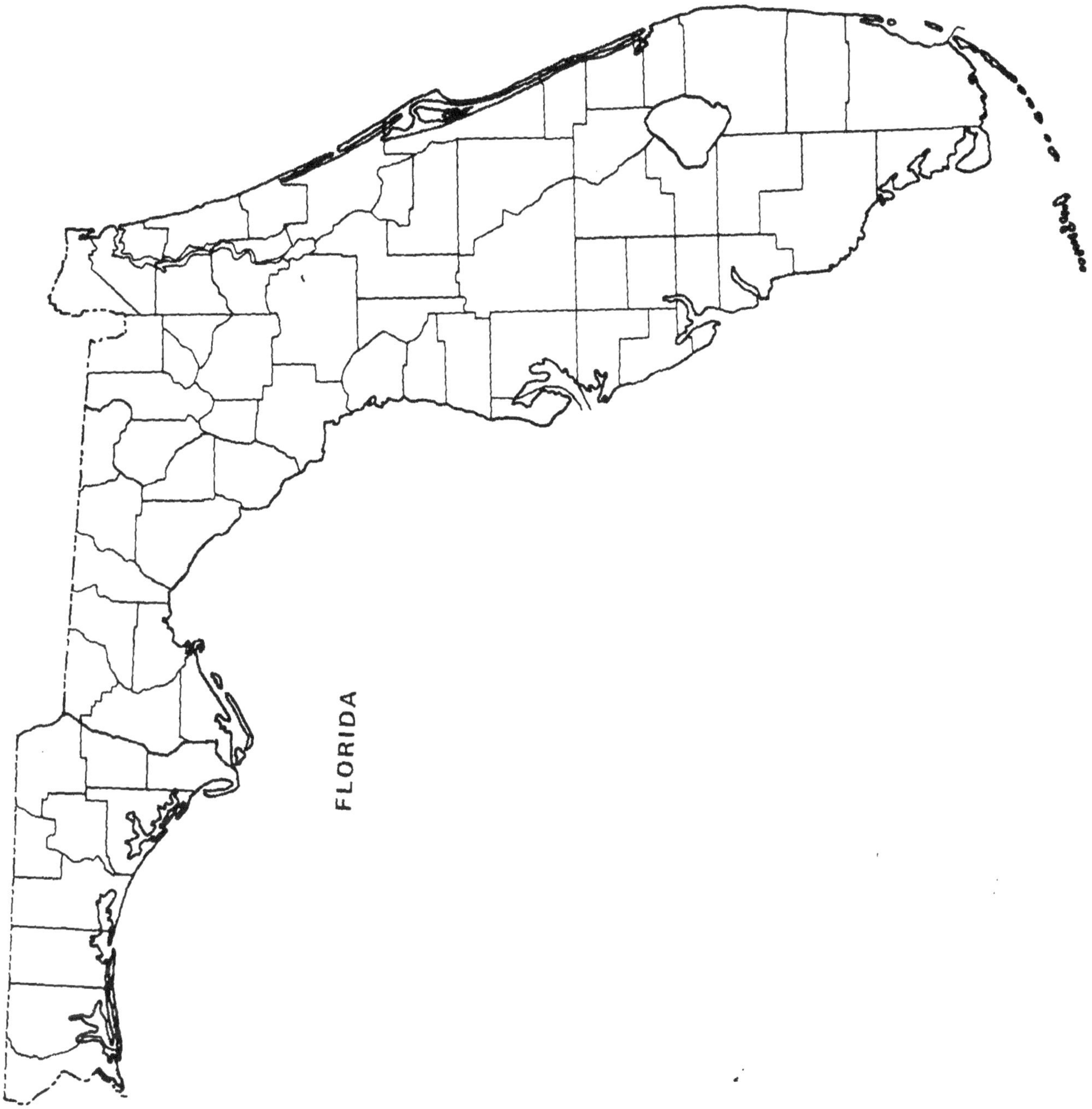
FLORIDA

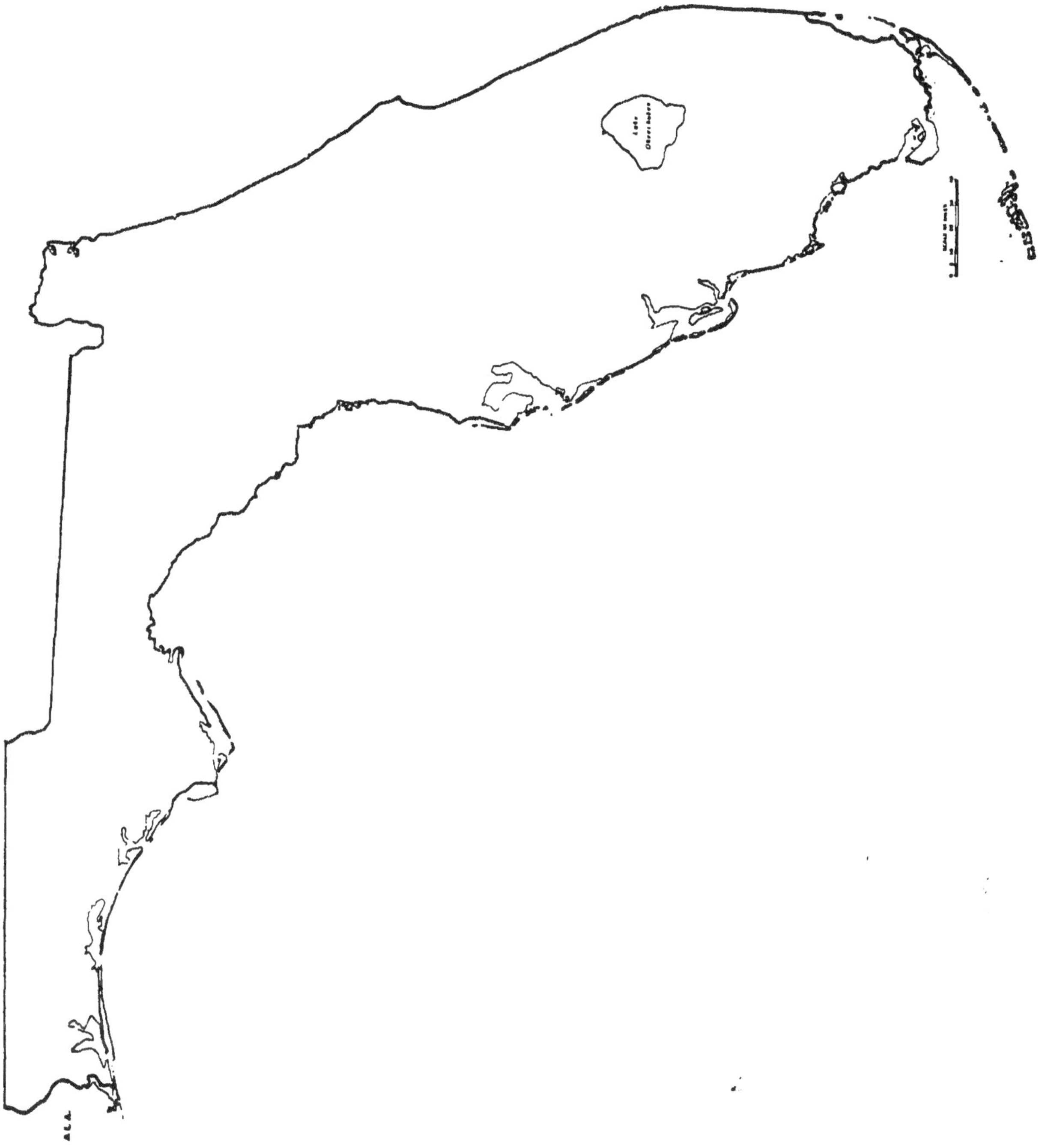

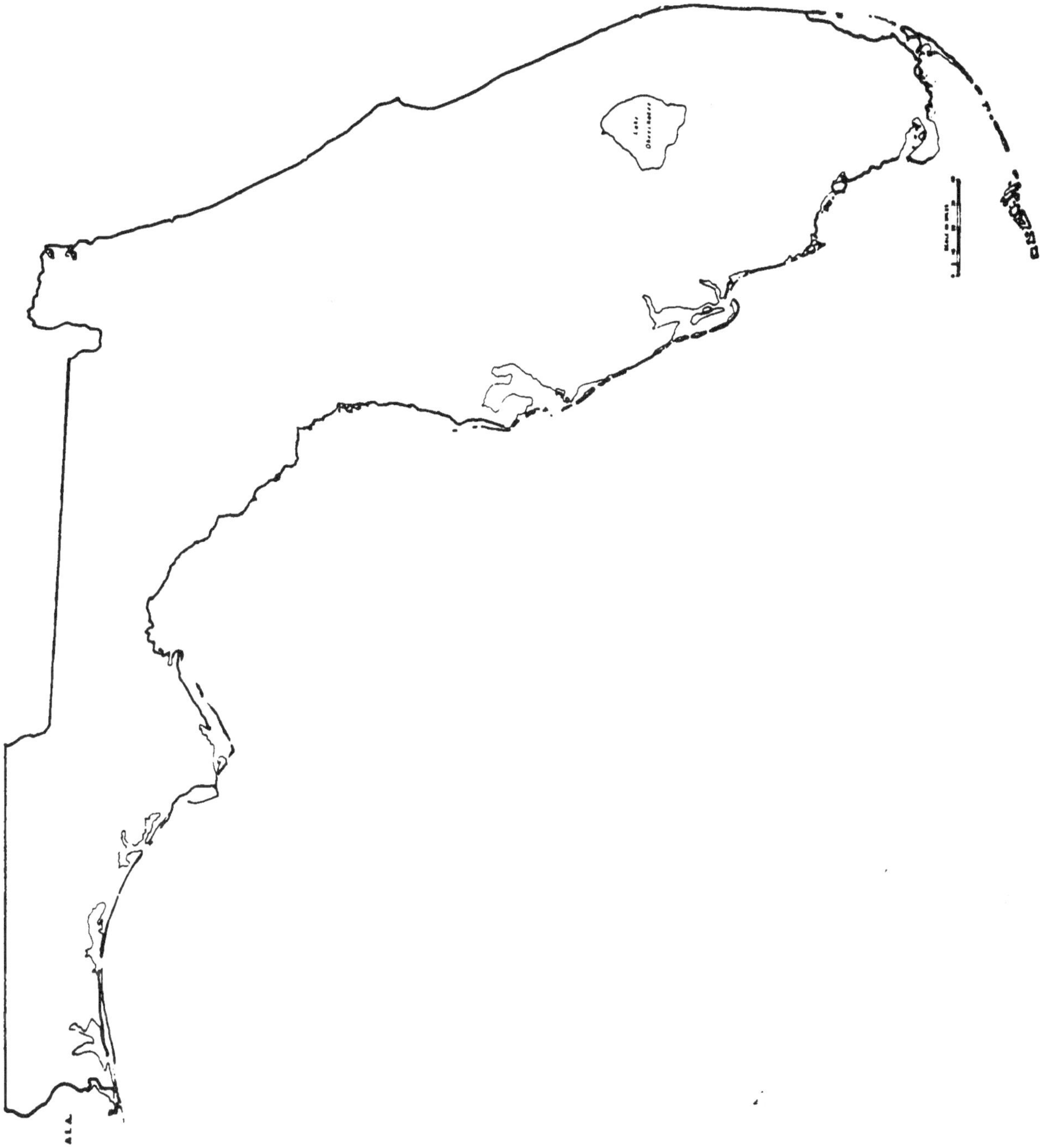
ALA.

Index*

*"il" following page number indicates an illustration, "m" a map, "n" a note, "t" a table.

www.ingramcontent.com/pod-product-compliance
Ingram Content Group UK Ltd.
Pitfield, Milton Keynes, MK11 3LW, UK
UKHW051509290926
15471UKWH00008B/1510

9 780787 248499